Groundwater Pollution Risk Control
from an Industrial Economics Perspective

Huan Huan · Jianwei Xu
Jinsheng Wang · Beidou Xi

Groundwater Pollution Risk Control from an Industrial Economics Perspective

A Case Study on the Jilin Section of the Songhua River

Huan Huan
Chinese Research Academy
  of Environmental Sciences
Beijing
China

Jianwei Xu
Institute of Industrial
  and Technological Economics
National Development and Reform
  Commission
Beijing
China

Jinsheng Wang
Beijing Normal University
Beijing
China

Beidou Xi
Chinese Research Academy
  of Environmental Sciences
Beijing
China

ISBN 978-981-10-7705-0   ISBN 978-981-10-7706-7 (eBook)
https://doi.org/10.1007/978-981-10-7706-7

Library of Congress Control Number: 2018937337

© Springer Nature Singapore Pte Ltd. 2018
This work is subject to copyright. All rights are reserved by the Publisher, whether the whole or part of the material is concerned, specifically the rights of translation, reprinting, reuse of illustrations, recitation, broadcasting, reproduction on microfilms or in any other physical way, and transmission or information storage and retrieval, electronic adaptation, computer software, or by similar or dissimilar methodology now known or hereafter developed.
The use of general descriptive names, registered names, trademarks, service marks, etc. in this publication does not imply, even in the absence of a specific statement, that such names are exempt from the relevant protective laws and regulations and therefore free for general use.
The publisher, the authors and the editors are safe to assume that the advice and information in this book are believed to be true and accurate at the date of publication. Neither the publisher nor the authors or the editors give a warranty, express or implied, with respect to the material contained herein or for any errors or omissions that may have been made. The publisher remains neutral with regard to jurisdictional claims in published maps and institutional affiliations.

Printed on acid-free paper

This Springer imprint is published by the registered company Springer Nature Singapore Pte Ltd. part of Springer Nature
The registered company address is: 152 Beach Road, #21-01/04 Gateway East, Singapore 189721, Singapore

Acknowledgements

This study was supported by National Natural Science Foundation of China (Grant No. 41602260) and Beijing Natural Science Foundation (Grant No. 8164066). The authors would like to thank the Environmental Geology Monitoring Station of Jilin Province, China, which provided the necessary data to conduct this study.

Contents

1 Introduction 1
  1.1 Research Background and Significance 1
      1.1.1 Background 1
      1.1.2 Significance 2
  1.2 Research Progress at Home and Abroad 3
      1.2.1 Groundwater Pollution Risk Assessment 3
      1.2.2 Relationship Between Industrial Economics
            and Groundwater Pollution 7
  1.3 Research Objective, Content and Technical Roadmap 9
      1.3.1 Objective 9
      1.3.2 Content 9
      1.3.3 Technical Roadmap 10
  References 11

2 Industrial and Economic Analysis of Groundwater Pollution ... 15
  2.1 Mechanism of Interaction Between Industrial Structure
      and Groundwater Pollution 15
  2.2 Water Pollution Characteristics of Different Sectors 17
  References 24

**3 Natural Circumstance and Industrial Economy of the Study
Area** 25
  3.1 Physical Geography 25
      3.1.1 Geographical Location 25
      3.1.2 Hydro-Meteorological Conditions 25
      3.1.3 Soil and Vegetation 26
  3.2 Geological Conditions 27
      3.2.1 Terrain and Landform 27
      3.2.2 Stratigraphy and Structure 27
      3.2.3 Hydrogeological Conditions 30

3.3	Groundwater Exploitation and Utilization	32
3.4	Industrial Characteristics of Jilin City	34
	3.4.1 Heavy Reliance on Resources	34
	3.4.2 Clear Dominance of Heavy Chemical Industry	35
	3.4.3 Relatively Low Technical Level	38
	3.4.4 Significant Spatial Variation of Industry	39
References		42

4 Groundwater Pollution Characteristics and Source Apportionment ... 43
  4.1 Hydro-Chemical Characteristics ... 43
  4.2 Groundwater Quality Evaluation ... 47
  4.3 Groundwater Pollution Assessment ... 48
  4.4 Groundwater Pollution Source Apportionment ... 54
    4.4.1 Groundwater Pollution Sources and Land Use Types ... 55
    4.4.2 Groundwater Pollution Source Apportionment ... 59

5 Groundwater Pollution Risk Assessment ... 63
  5.1 Intrinsic Groundwater Vulnerability Assessment ... 63
  5.2 Pollution Load Assessment ... 65
    5.2.1 Selection of Typical Pollutants ... 67
    5.2.2 Load of Typical Pollutants ... 68
    5.2.3 Calculation of Pollutant Emissions ... 70
    5.2.4 Pollution Load Evaluation Results ... 76
  5.3 Groundwater Function Evaluation ... 79
    5.3.1 Indicator System and Rating ... 79
    5.3.2 Indicator System Weight ... 87
    5.3.3 Results of Groundwater Function Evaluation ... 88
  5.4 Results and Verification ... 95
    5.4.1 Results ... 95
    5.4.2 Result Verification ... 97
  References ... 99

6 Economic Losses of Groundwater Pollution ... 101
  6.1 Aggregate Value of Groundwater Resources ... 101
  6.2 Economic Losses of Groundwater Pollution ... 105
  References ... 108

7 Groundwater Pollution Control Risk from the Perspective of Industrial Economics ... 109
  7.1 Thinking on Industrial Restructuring and Its Priorities ... 110
  7.2 Path for Industrial Restructuring ... 111
    7.2.1 Intensify Structural Adjustment of Industries of Different Types ... 111
    7.2.2 Push for Efficiency Improvements in Various Industrial Sectors ... 114

		7.2.3	Advance the Construction of Public Platforms and Infrastructure	116
	7.3		Summary and Implication	119
	References			120
8	**Conclusions**			121
	8.1		Groundwater Pollution Characteristics	121
	8.2		Groundwater Pollution Risk Assessment Results	121
	8.3		Aggregate Groundwater Value and Economic Losses Caused by Groundwater Pollution	122
	8.4		Groundwater Pollution Risk Control from the Perspective of Industrial Economics	122

Chapter 1
Introduction

Abstract Groundwater pollution risk assessment is the premise and foundation of groundwater pollution prevention and control. In order to make new breakthrough, groundwater pollution risk assessment should be combined with economic analysis to carry on environmental protection and management of groundwater. In Jilin city, the Songhua River, as the main source of water supply, has poor risk condition. The groundwater resources can play a leading role in emergency water supply. Industrial structure was gradually formed to be heavy industry-oriented in the study area. It's hard to improve the current condition of large consumption of resources, much discharge of pollutants, backward production technology and low utilization of resource in a short period of time. Therefore, it's significant to reduce the risk of groundwater pollution to carry on ecological construction, develop the appropriate industry development model and eventually realize the balance and sustainable development of resources, environment and economy on the base of industrial adjustment optimization study. The book adopts overlay index method to assess groundwater pollution risk on the condition of taking shallow groundwater as the research object. And then, the mechanism of the industrial structure and groundwater pollution is revealed by identifying high level of groundwater pollution risk and main pollution sources. Finally, the measures of groundwater pollution control are put forward from the perspective of industrial economics.

Keywords Groundwater pollution risk · Industrial economics
Interaction mechanism · Economic losses · Jilin · Songhua river
Groundwater pollution control

1.1 Research Background and Significance

1.1.1 Background

The world economy has entered a new era of development in the context of global climate change, economic globalization, informationalization and marketization. However, the rapid development of the global economy has undermined the environment and resources to a certain degree, and triggered a series of problems related to resource depletion and environmental degradation. The continued provision of resources and energy by the environment beyond its renewable capacity and carrying capacity will permanently damage environmental stability and resilience, thus restricting sustainable economic and social development. The essence of sustainable development is the coordinated development of population, resources, environment and economy. The economic and environmental studies of river basins have emerged as a hot spot and focus issue of sustainable development. As the world's economic corridors and economic centers largely concentrate in the river basins, high-pollution, high-consumption and high-emission (three-high) industries that dominate the upper reaches seriously pollute the local environment, especially the water environment.

The Jilin section of the Songhua River mainly flows through industrial-intensive and densely populated areas of Jilin Province. The long-established industrial structure dominated by heavy chemical industry presents high energy consumption and large pollutant emissions, and the leading industries are distributed along the river and in the upper reaches of the basin. Under the combined impact, water supply and demand contractions and water environmental problems have become increasingly prominent, mainly reflected in the growing water demand, decreased carrying capacity and continued deterioration of water environment, and environmental pollution emergencies (Yang 2013a). Since the Songhua River is closely related to groundwater hydraulics, the contaminated water, once merging into the riverside groundwater sources, will threaten the drinking water safety of urban and rural residents. Groundwater can be used as an important emergency water source by virtue of its strong self-purification ability, good water quality, and highly secured water supply. However, groundwater pollution is expanding in the Songhua River Basin, according to the long-term groundwater monitoring datasets of the Ministry of Land and Resources (MLR) and two national groundwater resource evaluations (conducted during 1981–1984 and 1999–2002). The overall groundwater quality has declined, with generally detected carcinogenic, teratogenic and mutagenic trace organic pollutants. In addition to the non-point source pollution of ammonia nitrogen, nitrite nitrogen and nitrate nitrogen caused by agricultural activities, the point source pollution of heavy metals and organic matter brought by accelerated urbanization and energy development also pose a serious threat to groundwater environmental safety. Therefore, the improvement of groundwater environmental quality has become a top priority, with a view to coordinated development of economic growth and environmental protection in the river basin. Given the three-high industrial structure in the upper reaches of the Songhua River Basin, the large-scale relocation of heavy industry is neither economical nor feasible. Industrial restructuring and optimization turns out to be an important way to harmonize economy and environment in the river basin.

1.1.2 Significance

Groundwater, as a major resource and a key environmental factor, is no doubt of increasing importance. Groundwater pollution, however, is escalating and the treatment and recovery is difficult and costly. To address groundwater pollution, we should "give priority to prevention and put more emphasis on prevention than control". Groundwater pollution risk assessment is an important prerequisite and basis for pollution prevention and control. It measures groundwater pollution risks and identifies and minimizes the acceptable risks, which will help policy makers and managers to develop groundwater protection strategies and policies (Stephen et al. 2002) and pave the theoretical foundation for further harmonious and sustainable development in economy and environment.

With the rapid social and economic development, the unreasonable use of groundwater by human beings has brought direct consequences (groundwater quality deterioration, decreased groundwater reserves, and etc.) and indirect consequences (environmental degradation, threat to human health, and etc.). In monetary terms, the consequence is embodied in the loss of groundwater value. Though basically meet the corresponding needs, groundwater pollution risk assessment should be combined with economic analysis to make new breakthroughs towards effective protection and management of water environment in a market economy (Liu 2007). The risk assessment results will be more acceptable and valuable in the regional planning, construction or management, if they make clear the economic losses of development without respect for the groundwater environment and put forward policies or measures to control the environmental risks.

In Jilin section of the Songhua River with poor risk resilience, abundant groundwater resources can play a leading role in emergency water supply while the Songhua River serves as the primary water source. As the region has a long-established industrial structure dominated by heavy industry, it is hard to address, in a short period of time, such problems as huge resource consumption, large pollutant emissions from production, backward production technologies, and low energy and resource utilization rate. Based on groundwater pollution risk assessment and through research into industrial restructuring and optimization, the establishment of a green production chain and introduction of an appropriate model for minimum-risk, eco-friendly industrial development will be of vital significance to the coordinated and sustainable development of resources, environment and economy in the region.

1.2 Research Progress at Home and Abroad

1.2.1 *Groundwater Pollution Risk Assessment*

(1) Concept and content of groundwater pollution risk

Groundwater pollution risk assessment can be traced back to the groundwater vulnerability assessment brought forward by French scholar Marjat in the 1960s, and has evolved from intrinsic groundwater vulnerability assessment and specific vulnerability assessment of overlay land uses. Hence, groundwater pollution risk is an extended concept of groundwater vulnerability (Zhang 2006). Depending on the objects of assessment, the risk assessment can be classified into three categories, i.e. health risk assessment based on human health, ecological risk assessment based on ecological environment, and pollution risk assessment based on groundwater functions (Liang 2009).

There has not yet been a unified definition of groundwater pollution risk. Finizio and Villa (2002) defined the groundwater pollution risk as the possibility of contamination to the groundwater environment. Morris and Foster (2006) considered it the possibility of an unacceptable level of contamination to groundwater in the aquifers due to human activities and believed the risk is a result of the interaction between aquifer vulnerability to pollution and pollution load caused by human activities. Civita and Maio M De (2006) referred the risk to the product of groundwater contamination possibility and expected damage to risk bearers. Zhou and Li (2008) argued that the groundwater pollution risk is pollution probability superposed on pollution consequences and represents the probability of unacceptable pollution to aquifer groundwater caused by human activities.

According to the source—path—receptor analysis, groundwater pollution risk assessment covers three levels: (a) instinct groundwater vulnerability, reflecting mainly the self-purification capacity of groundwater systems and partially, the speed and quality of pollutants that reach the aquifer; (b) pollution load; and (c) expected harm of groundwater pollution system, i.e. level of risk acceptable for receptors (human health, ecological environment, society and economy, and etc.) (Shen and Li 2010; Teng et al. 2012).

(2) Objects and methods of groundwater pollution risk assessment

Groundwater pollution risk can be assessed at the scale of river basins (Dimitriou et al. 2008), water source areas (Nobre el al. 2007), cities (Chisala et al. 2007), sites (Li et al. 2012), and etc. The assessment may be focused on different typical pollutants from various sources (Wang et al. 2012) or specific to certain pollutant or pollutant type, such as nitrates (Mario et al. 2014; Victor et al. 2014), methyl tert-butyl ether (MTBE) (Chisala et al. 2007), As (Zhang et al. 2012) and pesticide (Zhao and Pei 2012).

At present, the methods for groundwater contamination risk assessment include the empirical method (qualitative assessment), index overlay method, process simulation method, and statistical method (Shen and Guang 2010; Teng et al. 2012). Among them, the index overlay method is simple and practical, and therefore most widely used. Its common applications include the Hazard Ranking System (HRS) of the United States Environmental Protection Agency (EPA) (EPA 1992), Groundwater Vulnerability Scoring System (GVSS) (Hathhorn and Wubbena 1996) and the USGS Method of the United States Geological Survey (USGS) (Eimers et al. 2000), and the Susceptibility Index (SI) Method of Portuguese Geological Information Center (Diamantino et al. 2005).

The typical practice at home and abroad is to integrate pollution load factors into instinct groundwater vulnerability assessment and superimpose the results on groundwater value assessment using the overlay, cross table, matrix or HRS to obtain the groundwater pollution risk index and its distribution map (Teng et al. 2012).

A. Groundwater vulnerability assessment

Groundwater vulnerability assessment methods include the index overlay method, process simulation method, statistical method, fuzzy mathematics method and the synthesis of various methods. They can use such models as DRASTIC, GOD, AVI, SINTACS, ISIS, Legrand, SIGA and SEEPAGE, and when specific to karst aquifers, GLA, EPIK, PI, VULK, COP and LEA. Among them, the DRASTIC model of the index overlay method is the most widely used. The refined models by increasing or decreasing indicators are used to assess the vulnerability of concealed water and confined water throughout the United States, Canada, South Africa and the European Community, presenting quite reasonable results.

B. Pollution load risk assessment

Table 1.1 lists several common methods for pollution load risk assessment abroad. Among them, the Danger of Contamination Index (DCI) method is a simple and quick qualitative approach that adds animal production and other non-point sources to industrial activities. The Pollutant Origin and Its Surcharge Hydraulically (POSH) method classifies risks according to pollution sources and generated pollution loads. It is strongly practicable owning to the biggest advantage of moderate data demand and data availability.

C. Groundwater function assessment

Table 1.2 shows the common method for groundwater function assessment at home and abroad. These methods use groundwater value to represent groundwater functions. Groundwater value is made up of mining value and in situ value (US. NRC 1997) which from an economic point of view, can be understood as use value and non-use value. Most of the methods examine the use value through indicators of mining value in different aspects, but generally select groundwater quality for in situ value evaluation because the indicators are difficult to quantify in practice.

There are some studies on the protection of groundwater sources and maintenance of groundwater ecological functions abroad, but little work on the classification of groundwater functional areas and the principles, so the relevant information of reference value is limited. In 2003, China Geological Survey (CGS) deployed the work of assessing and zoning groundwater functions, marking the start of groundwater function assessment in the country. Issued by CGS in 2006, the *Technical Specifications for Assessing and Zoning Groundwater Functions* (GWI-D5) establishes the GIS-based functional assessment indicator system, which is suitable for the quaternary groundwater system of the northern plains, and becomes the work standards for groundwater function assessment and zoning (Sun and Li 2013).

Table 1.1 Pollution load risk assessment methods

Method	Description	Result property	Advantage	Limitation
Simple judgment (Zaporozec 2002)	Classifies contaminants into seven categories, i.e. nature, agriculture, forestry, life, solid waste, wastewater treatment, industry and mining, and water management failure, and grades the risk as high, medium or low based on experience	Qualitative	Simple and quick, and less data demand	Impact of human factors on indicator scores, and lack of regional comparison
DCI	Divides pollutants into four categories, i.e. industry, agriculture, animal husbandry and others, and further into different types according to the scale of these sectors	Qualitative	Easy to use and more detailed compared with the simple judgment	Lack of regional comparison
Detailed grading	Refines classification and lacks regional comparison	Quantitative	Avoids human subjectivity, and contributes to pollution source treatment and groundwater protection	Requires in-depth field surveys to obtain detailed information on a large number of sources of pollution
POSH	Classifies risks according to pollution sources and generated pollution loads	Qualitative and quantitative	Less data demand and strong operability	Lack of comparison between different types of sources
Priority-setting method (USEPA 1991; Harman et al. 2001)	Takes into account the possibility and severity of pollution	Quantitative	Clear classification system and risk comparison between different sources of pollution	Fails to cover all types of pollution sources and requires detailed investigation of pollution sources

(3) Problems of groundwater pollution risk assessment

First, the definition of groundwater pollution risk is not unified in existing studies, so in-depth discussion of its concept and connotation is needed. Second, the methods for evaluating pollution loads and groundwater functions are not yet perfect with incomplete and mostly qualitative indicators. Third, surface pollutants,

Table 1.2 Groundwater function assessment methods

Proposer	Description	Indicator
US.NRC (1997)	Grades risk based on the use standards of water rather than quality standards	Groundwater use type, wastewater discharge type
Parsons (1995)	Grades risk using the matrix of aquifer types and user-defined variables	Aquifer classification
Ducci (1999)	Suggests relating to the size of water supply area	Population supported
Li et al. (2010)	Determines according to groundwater usage and corresponding requirements for water supply assurance rate	None
Jiang et al. (2010)	Evaluate the value according to groundwater quantity and quality and significance for water supply	Single-well water yield, water quality, water source protection area
Zhang Lijun (2006)	Considers ecological/health services and socio-economic services of groundwater, including water quality and quantity	Comprehensive groundwater quality index, recoverable groundwater, per capita groundwater share, and groundwater supply rate for production

unsaturated zones and aquifers are not considered as a whole process, while the migration and transformation of pollutants in the unsaturated zones and aquifers is neglected. Fourth, the groundwater pollution risk assessment is mainly focused on typical pollutants rather than multiple pollutants. In addition, the assessment results are not validated.

1.2.2 Relationship Between Industrial Economics and Groundwater Pollution

The research on the relationship between environmental pollution and economic development has started early, represented by the environmental Kuznets curve (EKC) proposed by Grossman and Krueger (1991). In the field of groundwater pollution, one of the focal points of studies is groundwater resources, groundwater pollution and the relationship with economic development. Groundwater resources are of ecological, economic and social significance (Vrba and Shu 1991) which often support mutually. Yang and Wan (2007) constructed an indicator system to preliminarily evaluate the role of groundwater resources in supporting economic and social development in provinces. Wang et al. (2012), Feng and Li (2006) built a quantitative evaluation model to analyze the interaction between groundwater environment and socio-economic development. Tapio (2005) believed it necessary to classify pollutants as the relationship between economic growth and environmental pollution is very vulnerable to the sample selection and research methodology. Nahar et al. (2008) concluded the relationship between per capita income and groundwater water quality based on the sampling arsenic contamination survey of groundwater wells in rural areas of Bangladesh. Zhong and Wei (2011) found that per capita gross domestic product (GDP) and urbanization rate are significantly correlated with groundwater quality in Wenyu River Basin. Zhang and Wei (2006) also found that groundwater pollution evaluation results match economic development and there is a positive correlation between economic development and groundwater pollution.

Economic losses caused by groundwater pollution are another focus of research on the relationship between environmental pollution and economic development. In this regard, there have been a large number of research findings owning to studies conducted early at home and abroad. Guo and Wei (1999) described the basic concepts and principles of economic evaluation of groundwater pollution, and proposed the depletion impairment theory according to the relationship among function, value and pollution of groundwater resources. Tan and Zheng (2003) believed that the calculation of economic losses caused by groundwater pollution should consider not only the self-purification capacity of groundwater bodies, but also such factors as the cumulative effect of pollutants on the environment, and on this basis, built an economic evaluation model for groundwater pollution. The economic losses can be calculated according to polluted water bodies or pollution-effected objects. In the former scenario, the commonly used methods include the recovery cost method, fuzzy mathematics method, and opportunity cost method. Liu and Ye et al. (2006) measured the economic losses caused by groundwater pollution based on concentration -value loss factor analysis through a case study of Hutuohe groundwater source in Shijiazhuang. In the later scenario, the calculation considers the losses in the aspects of human health, agriculture, industry and fisheries, using the human capital method and the market value method (Wang and Wang 2006). According to the characteristics of water pollution and agricultural economy, Zhang and Cao (2008), Han and Wang (2009) applied the shadow price method to calculate the market value of agricultural crops and estimate the environmental costs caused by irrigation with polluted water.

Groundwater pollution is essentially an integrated matter of structural problems. Simple environmental and economic policies (including pollution and emission permits, pollution charges, pollution taxes, ecological compensation, and pollution compensation) can not effectively control the risk of groundwater pollution. The fundamental solution to groundwater pollution should rest on the analysis of industrial structure mechanisms of groundwater pollution. The logical basis for the relationship between industrial structure and environmental pollution is different pollution characteristics among sectors. Given this, Liu and Zhao (2012) introduced the concept of industrial environment (pollution) intensity, and divided the economy into water-pollution-intensive sectors, air-pollution-intensive sectors, and solid-waste-intensive sectors. There are also studies on the causes and sources of groundwater contamination from the industrial point of view, such as Yang (2013a) and so on. Groundwater pollution can be divided into industrial source pollution, agricultural source pollution, domestic source pollution and natural source pollution (Peng and Zhang 2005), but this classification is obviously too broad and rough. Furthermore, Yang (2013b) constructed a coupling mechanism of industrial system and environmental system, quantitatively measured the interaction between industrial structure and industrial wastewater discharge, and put forward policy recommendations on internal economic growth. In general, the research on risk control of groundwater pollution from the perspective of industrial economies remains weak. The literature on the relationship between industrial structure evolution and groundwater pollutant discharge based on the pollution characteristics of different sectors are still inadequate. Industrial economics from the structuralist perspective can probe into the root causes of different groundwater pollution and play a crucial role in pollution prevention and control and risk control. Hence, such studies are expected to direct towards the integration of environmental science and industrial economics.

1.3 Research Objective, Content and Technical Roadmap

1.3.1 Objective

Through the assessment of intrinsic groundwater vulnerability, pollution loads and groundwater functions and the analysis of groundwater pollution characteristics, the study will (1) reveal the risk distribution characteristics of groundwater pollution and identify the dominant factors of high-risk areas, which will lay a theoretical basis for risk control and pollution control measures; and (2) figure out the mechanism of interaction between industrial structure and groundwater pollution based on pollution characteristics of different sectors, and bring forward initiatives to adjust and optimize the industrial structure, so that economic development in drainage basins and regions can be sustained in harmony with the groundwater environment.

1.3.2 Content

The study will encompass:

(1) Evaluation of groundwater quality and groundwater pollution, and identification of groundwater pollution characteristics and sources.
(2) Groundwater pollution risk assessment, covering a variety of pollution sources and typical pollutants.

Intrinsic groundwater vulnerability, pollution loads and groundwater functions will be evaluated according to the established indicator systems. Based on the distribution of intrinsic groundwater vulnerability and groundwater functions, the groundwater protection urgency will be mapped to represent groundwater pollution consequences and the pollution hazard ranking used to represent groundwater pollution possibility. The two are combined to map the groundwater pollution risk.

(3) Calculation of economic losses caused by groundwater pollution.
(4) Recommendation of industrial optimization measures from the perspective of industrial economics according to high-risk areas delineated based on groundwater pollution risk assessment and reflecting groundwater pollution sources. These measures are expected to contribute to the sustainable development of regional economy while mitigating the risk of groundwater pollution.

The mechanism of interaction between groundwater pollution and industrial structure will be analyzed, and then the characteristics of groundwater pollution in different sectors revealed. With consideration to industrial and economic characteristics of the study area, industrial restructuring will be suggested based on groundwater pollution risk control, to introduce a new model of industrial development.

1.3.3 Technical Roadmap

The technical roadmap of the study is drawn, as shown in Fig. 1.1, according to the research purpose and content.

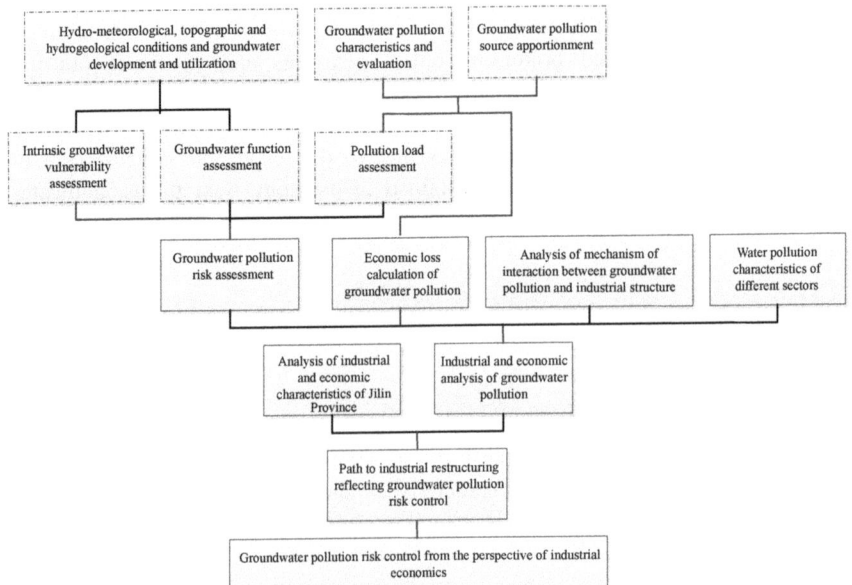

Fig. 1.1 Technical roadmap

References

Chisala BN, Tait NG, Lerner DN. Evaluating the risks of methyl tertiary butyl ether (MTBE) pollution of urban groundwater. J Contam Hydrol. 2007;91:128–45.

Civita MV, Maio M De. Assessment groundwater contamination risk using Arcinfo via GRID function. http://proceedings.esri.com/liberary/userconf/proc97/proc97/to600/pap591/p591.htm, 2006-05-02.

Diamantino C, Henriques MJ, Oliveira MM, et al. Methodologies for pollution risk assessment of water resources systems. In: Proceedings of the fourth interceltic colloquium on hydrology and management of water resources, Guimaraes, Portugal;2005.

Dimitriou E, Karaouzas I, Sarantakos K, et al. Groundwater risk assessment at a heavily industrialised catchment and the associated impacts on a peri-urban wetland. J Environ Manag. 2008;88(3):526–38.

Ducci D. GIS techniques for mapping groundwater contamination risk. Nat Hazards. 1999;20:279–94.

Eimers L, Weaver J, Terziotti S, et al. Methods of rating unsaturated zone and watershed characteristics of public water supplies in North Carlina. Raleigh, North Carolina: USGS; 2000.

Feng P, Li J. Model for analyzing coordinated groundwater and socio-economic development and its application. Math Practice Theory. 2006;4:61–6.

Finizio A, Villa S. Environment risk assessment for pesticides: A tool for decision-makers. Environ Impact Assess Rev. 2002;22(3):235–48.

Grossman GM, Krneger AB. Environmental impacts of a north american free trade agreement. In: National bureau of economic research working paper;1991.

Guo ZX, Wei YF. A study on methods for economic evaluation of groundwater sources and environmental pollution. In: Outstanding Proceedings of chinese hydraulic engineering society in 1999;1999.

Han MQ, Wang LG. Exploration to economic losses caused by water environmental pollution and agricultural losses. South-to-North Water Transf Water Sci Technol. 2009;7(1):76–8.

Harman WA, Allanand CJ, Forsythe RD. Assessment of potential groundwater contamination sources in a wellhead protection area. J Environ Manag. 2001;62(3):271–82.

Hathhorn W, Wubbena T. Site vulnerability assessment for wellhead protection planning. J Hydrol Eng. 1996;4:152–60.

Jiang J, Dong DW, Yang GN, et al. Groundwater pollution risk assessment of Haidian district in Beijing. Anal Study. 2010;5(2):44–8.

Li Y, Li JH, Chen SS, Diao WH. Establishing indices for groundwater contamination risk assessment in the vicinity of hazardous waste landfills in China. Environ Pollut. 2012;165:77–90.

Li RZ, Wang MW, Jin J. Fuzzy multiple attribute analysis of groundwater environmental risk. Sci Geogr Sin. 2010;3(2):229–35.

Liang J. A study of solute transport and pollution risk of groundwater based on the uncertainty theory. Changsha: Hunan University; 2009.

Liu CL. Economic evaluation of geological environmental risk of cities. Shijiazhuang: Chinese Academy of Geological Sciences; 2007.

Liu CL, Ye H, Dong H, et al. Application of "Concentration-Value Loss Factor" to Assess Economic Loss of Groundwater Pollution in Hutuo-River of Shijiazhuang [J]. Resources Science, 2006;28(6): 2–9.

Liu H, Zhao JF. Decoupling heterogeneity of pollution-intensive industries and industrial transformation in China. China Popul Resour Environ. 2012;22(4):150–5.

Mario CO, Juan AL, Victor RG, et al. Categorical indicator kriging for assessing the risk of groundwater nitrate pollution: The case of Vega de Granada aquifer (SE Spain). Sci Total Environ, 2014;470–471 (1):229–239.

Morris B, Foster S. Assessment of groundwater pollution risk. http://www.inweb18.worldbank.org/essd/essd.nsf, 2006-05-06.

Nobre RCM, Rotunno OC, Mansur WJ, et al. Groundwater vulnerability and risk mapping using GIS, modeling and a fuzzy logic tool. J Contam Hydrol. 2007;94(3–4):277–92.

Nurun N, Faisal HM, Delawer H. Health and socioeconomic effect of groundwater arsenic contamination in rural bangladesh new evidence from field surveys. J Environ Health. 2008;70(9):42–7.

Parsons R. A South African aquifer system management classification. Water Research Commission;1995.

Peng WQ, Zhang XW. Theories and methods of modern water quality Assessment. Beijing: Chemical Industry Press; 2005.

Shen LN, Li GH. Research on groundwater pollution risk zoning methods. Acta Sci Circum. 2010;31(4):918–23.

Stephen F, Ricardo H, Daniel G, et al. Groundwater quality protection, a guide for water utilities, municipal authorities, and environment agencies. Washington D C: The World Bank; 2002.

Sun CZ, Li XM. Groundwater function evaluation of the Liaohe River Plain based on ArcGIS. Sci Geogr Sin. 2013;33(2):174–80.

Tan WB, Zheng M. Exploration to economic evaluation model for groundwater pollution. In: Resources environment industry—Proceedings of 2003 annual conference of china society on economics of geology & mineral resources. September 2003.

Tapio P. Towards a theory of decoupling: degrees of decoupling in the EU and the case of road traffic in Finland between 1970 and 2001. Transport Policy 2005;12(2):137–151

Teng YG, Su J, Zhai YZ, et al. A review of index overlay analysis in groundwater pollution risk assessment. Adv Earth Sci. 2012;27(10):1140–7.

U.S Environmental Protection Agency. The hazard ranking system guidance manual. Washington DC: Hazardous site evaluation division office of solid waste and emergency response, U.S. Environmental Protection Agency;1992.

U.S Environmental Protection Agency. Managing groundwater contamination sources in wellhead protection areas- A priority setting approach. USA. EPA;1991.

U.S. National Research Council. Valuing Groundwater economic concepts and approaches. Washington DC, USA: National Academy Press;1997.

Victor RG, Maria PM, Maria JGS, et al. Predictive modeling of groundwater nitrate pollution using Random Forest and multisource variables related to intrinsic and specific vulnerability: A case study in an agricultural setting (Southern Spain). Sci Total Environ. 2014;476–477(1):189–206.

Vrba J, Shu LC. Ecological, economic and social significance of groundwater protection. World Geol. 1991;1:170–1.

Wang JJ, He JT, Chen HH. Assessment of groundwater contamination risk using hazard quantification, a modified DRASTIC model and groundwater value, Beijing Plain, China. Sci Total Environ. 2012;432(15):216–26.

Wang Y, Wang C. Economic losses caused by water pollution in Shandong Province. China Popul Resour Environ. 2006;16(2):83–7.

Yang LH. A study of economic and environmental effects of Songhua River Basin (Jilin Province) and industrial space organization. Beijing: University of Chinese Academy of Sciences;2013a.

Yang JF, Wan SQ. Assessment of the support of groundwater resources for regional economic and social development in China. Res Sci. 2007;5:97–104.

Yang KL. Research on groundwater pollution prevention and control in Hubei Province. J Yangtze University (Nat Sci Edn) 2013b;(34):19–23.

Zaporozec A. Groundwater contamination inventory-A methodological guide. IHP-VI: UNESCO; 2002.

Zhang LJ. A summary of research progress on groundwater vulnerability and risk evaluation. Hydrogeol Eng Geol. 2006;6:113–8.

References

Zhang Q, Luis Rodríguez-Lado C, Annette J, Xue HB, Shi JB, Zheng QM, Sun GF. Predicting the risk of arsenic contaminated groundwater in Shanxi Province, Northern China. Environ Pollut. 2012;165:118–23.

Zhang JL, Wei XJ. AHP based comprehensive groundwater evaluation. J Hunan University (Nat Sci Edn). 2006;5:136–40.

Zhang LY, Cao SL. Evaluation of economic losses caused by groundwater environmental damage. Groundwater. 2008;1:4–8.

Zhao YY, Pei YS. Risk evaluation of groundwater pollution by pesticides in China: a short review. Procedia Environ Sci. 2012;13:1739–47.

Zhong W, Wei YS. Analysis of socio-economic development impact on groundwater in the yongding river basin (Beijing section) and wenyu river basin. Acta Sci Circum. 2011;9:1826–34.

Zhou YX, Li WP. Groundwater quality monitoring and assessment. Hydrogeol Eng Geol. 2008;35(1):1–11.

Chapter 2
Industrial and Economic Analysis of Groundwater Pollution

Abstract Since that environmental resources are valuable, the developed countries and regions around the world are reforming the existing national economic account system to include economic losses of environmental resources, so as to truly reflect the development of the national economy. The integration of economic losses caused by groundwater depletion, including groundwater pollution, into economic construction costs, is of great significance to environmental protection, resource conservation and national economic development. The economic losses caused by groundwater pollution are calculated based on the results of groundwater pollution risk assessment. The measurement more intuitively provides a dynamic and scientific basis for groundwater environmental protection and governance, which supports the establishment and implementation of groundwater pollution prevention and control system and the sustainability of economic development. This chapter probes into the mechanism of interaction between industrial structure and groundwater pollution and reveals the water pollution characteristics of different sectors.

Keywords Groundwater pollution · Industrial structure · Interaction mechanism Pollution characteristics · Different sectors

2.1 Mechanism of Interaction Between Industrial Structure and Groundwater Pollution

Industrial structure, as a resource converter, also functions on the input and output conversion of groundwater resources. Industrial development supplies materials and service products that meet the needs of economic and social development. Meanwhile, it discharges wastewater and other pollutants, which are by-products of the production process, to the ecological environment. In other words, production corresponds to economic growth, while wastewater discharge corresponds to groundwater capacity occupation. Generally speaking, with faster economic growth and larger economic scale, the demand of life will be satisfied to a higher degree, but wastewater discharge in the production process is constrained by environmental capacity. Environmental problems are caused by human activities that exceed the environmental carrying capacity. In the process of economic growth for satisfaction purpose, the discharge of wastewater beyond the environmental capacity will undermine the harmony between economic growth and environmental capacity, putting the entire economic and environmental systems in a high risk. At this point, in order to avoid system problem or collapse, we must control the environmental risks and furthermore, adjust the industrial structure. Industrial restructuring can be achieved by changing the proportion of industries, such as reducing the proportion of polluting industries in the economy, or by improving production efficiency, such as cutting emissions per unit of economic growth through technological progress. Ultimately, structural changes or efficiency improvements will reshape the industrial structure that controls wastewater within the scope of environmental capacity while meeting the needs of economic growth. At this time, the entire economic and environmental systems will be in a low-risk state. The mechanism of interaction between industrial structure and groundwater pollution is as shown in Fig. 2.1.

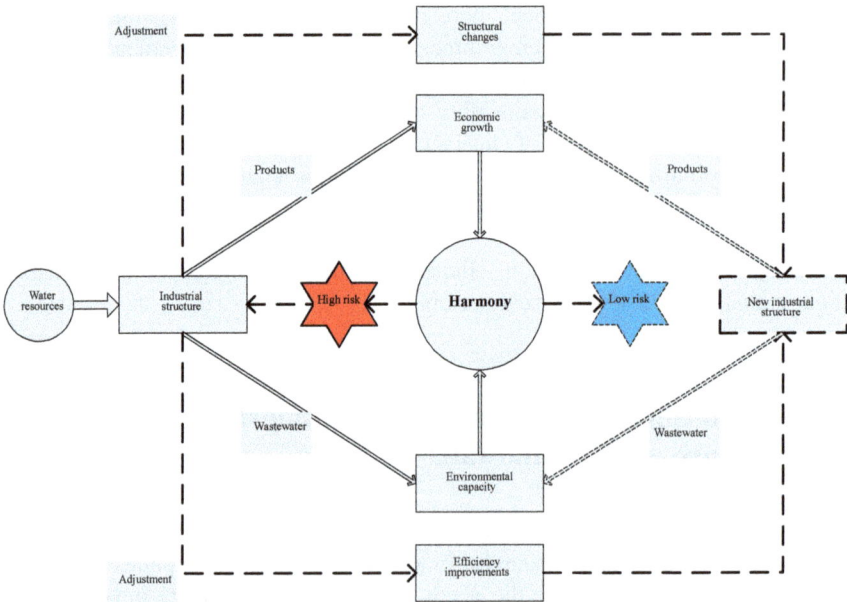

Fig. 2.1 Diagram of the mechanism of interaction between industrial structure and groundwater pollution Note: The solid line represents the process line under the initial industrial structure, and the dotted line represents the reaction line for adjustment according to the environmental risk. The figure is reprinted from Huan et al. (2016), with permission from Research On Development

2.2 Water Pollution Characteristics of Different Sectors

Pollution characteristics vary among different sectors. Groundwater pollution can be roughly analyzed in two aspects: (a) discharge intensity of output value, which refers to the amount of wastewater generated per unit of output value or scale of wastewater discharge; (b) pollution control difficulty, measured by the costs of treatment per unit of wastewater discharge, which is related to wastewater composition. Taking into account the two factors, the wastewater pollution intensity index is set to be wastewater treatment costs per 10,000 yuan of output value. Table 2.1 summarizes the wastewater pollution characteristics of different sectors in 2011. The method of quartering is used to grade three indicators, namely discharge intensity of output value (ton/10,000 yuan), pollution control difficulty (yuan/ton) and wastewater pollution intensity (yuan/10,000 yuan), and the results are as shown in Table 2.2.

According to pollution control difficulty, the industrial sectors can be divided into four groups, as shown in Table 2.3. High pollution intensity is mainly seen in water-consuming chemical industry, such as textiles, chemical fiber manufacturing, chemical and chemical product manufacturing, pharmaceutical manufacturing and paper making. Industries with relative high pollution intensity are scattered, covering the chemical industry, such as oil processing, coking and nuclear fuel processing, rubber products, the water-consuming light industry, such as leather, fur, feather (down) and their products, agricultural and sideline food processing, and also equipment manufacturing difficult to control pollution, such as manufacturing of general equipment, communications equipment, computers and other electronic equipment, metal products. Industries with low moderate and low pollution intensity mostly have short industrial chains and do not rely on other industries of raw materials. They include water production and supply, electricity and heat production and supply industry, gas production and supply, printing and recording media replication, and wood processing and wood, bamboo, rattan, palm and grass products, as well as water-efficient manufacturing, such as manufacturing of transportation equipment, electrical machinery and equipment, instruments and cultural office machinery, and special equipment.

According to *China Environmental Statistical Yearbook* (2012), four sectors generated the most wastewater, including paper making and paper products, chemical materials and chemical product manufacturing, textiles, electricity and heat production and supply. In 2011, they accounted for 50.3% of national total industrial wastewater discharge. The chemical oxygen demand (COD) emission intensity was much higher in the sectors of paper making and paper products, chemical fiber manufacturing, beverage manufacturing, agricultural and sideline food processing, and even up to 61.40 tons per 100 million yuan in the sector of paper making and paper products. The highest ammonia nitrogen emission

Table 2.1 Wastewater pollution characteristics of Chinese sectors in 2011

Item	Operating costs of wastewater treatment facilities (10,000 yuan)[1]	Wastewater treatment capacity (10,000 tons)[1]	Wastewater discharge (10,000 tons)[1]	Total output value (10,000 yuan)[1]	Discharge intensity of output value (tons/10,000 yuan)[2]	Pollution control difficulty (yuan/ton)[3]	Wastewater pollution intensity (yuan/10,000 yuan)[4]
Total	7,321,459	5,805,511	2,129,036	844,269	2.52	1.26	3.18
Coal mining and washing	288,374	185,121	143,493	28,920	4.96	1.56	7.73
Oil and gas extraction	229,311	87,903	8,172	12,889	0.63	2.61	1.65
Ferrous metal mining	136,894	287,995	22,643	7,904	2.86	0.48	1.36
Non-ferrous metal mining	411,493	162,009	51,181	5,035	10.17	2.54	25.82
Non-metallic mineral mining	13,226	10,894	6,191	3,848	1.61	1.21	1.95
Other mining	297	316	247	17	14.76	0.94	13.87
Agricultural and sideline food processing	175,065	127,644	138,116	44,126	3.13	1.37	4.29
Food manufacturing	124,814	45,946	51,950	14,047	3.70	2.72	10.05
Beverage manufacturing	193,869	62,391	71,664	11,835	6.06	3.11	18.82
Tobacco products	5,061	2,657	2,090	6,806	0.31	1.90	0.58
Textiles	547,962	205,859	240,802	32,653	7.37	2.66	19.63
Textile apparel, footwear and hat manufacturing	220,181	16,136	19,878	13,538	1.47	13.65	20.04
Leather, fur, feather (down) and their products	57,293	19,760	25,785	8,928	2.89	2.90	8.37
Wood processing and wood, bamboo, rattan, palm and grass products	6,231	2,304	3,522	9,002	0.39	2.70	1.06

(continued)

2.2 Water Pollution Characteristics of Different Sectors

Table 2.1 (continued)

Item	Operating costs of wastewater treatment facilities (10,000 yuan)[1]	Wastewater treatment capacity (10,000 tons)[1]	Wastewater discharge (10,000 tons)[1]	Total output value (10,000 yuan)[1]	Discharge intensity of output value (tons/10,000 yuan)[2]	Pollution control difficulty (yuan/ton)[3]	Wastewater pollution intensity (yuan/10,000 yuan)[4]
Furniture manufacturing	1,723	502	735	5,090	0.14	3.43	0.50
Paper making and paper products	610,290	550,237	382,265	12,080	31.65	1.11	35.10
Printing and recording media replication	4,569	978	1,303	3,861	0.34	4.67	1.58
Cultural, educational and sporting goods manufacturing	5,020	1,723	1,937	3,212	0.60	2.91	1.76
Oil processing, coking and nuclear fuel processing	508,390	199,200	79,587	36,889	2.16	2.55	5.51
Chemical and chemical product manufacturing	997,769	524,258	288,331	60,825	4.74	1.90	9.02
Pharmaceutical manufacturing	146,961	43,057	48,586	14,942	3.25	3.41	11.10
Chemical fiber manufacturing	90,074	38,184	41,428	6,674	6.21	2.36	14.64
Rubber product manufacturing	23,317	10,372	12,155	7,331	1.66	2.25	3.73
Non-metallic mineral product manufacturing	131,412	61,778	26,075	40,180	0.65	2.13	1.38
Ferrous metal smelting and calendering	1,208,045	2,461,163	121,037	64,067	1.89	0.49	0.93

(continued)

Table 2.1 (continued)

Item	Operating costs of wastewater treatment facilities (10,000 yuan)[1]	Wastewater treatment capacity (10,000 tons)[1]	Wastewater discharge (10,000 tons)[1]	Total output value (10,000 yuan)[1]	Discharge intensity of output value (tons/10,000 yuan)[2]	Pollution control difficulty (yuan/ton)[3]	Wastewater pollution intensity (yuan/10,000 yuan)[4]
Non-ferrous metal smelting and calendering	138,993	164,555	33,545	35,907	0.93	0.84	0.79
Metal product manufacturing	229,906	50,075	29,912	23,351	1.28	4.59	5.88
General equipment manufacturing	185,318	8,148	11,973	40,993	0.29	22.74	6.64
Special equipment manufacturing	10,384	4,149	6,454	26,149	0.25	2.50	0.62
Transportation equipment manufacturing	78,007	23,004	28,422	63,251	0.45	3.39	1.52
Electrical machinery and equipment manufacturing	35,425	8,193	9,631	51,426	0.19	4.32	0.81
Manufacturing of communications equipment, computers and other equipment	205,835	46,152	44,961	63,796	0.70	4.46	3.14
Manufacturing of instruments and cultural and office machinery	7,495	2,491	2,242	7,633	0.29	3.01	0.88
Handicrafts and other manufacturing	10,065	3,655	2,069	7,190	0.29	2.75	0.79

(continued)

2.2 Water Pollution Characteristics of Different Sectors

Table 2.1 (continued)

Item	Operating costs of wastewater treatment facilities (10,000 yuan)[1]	Wastewater treatment capacity (10,000 tons)[1]	Wastewater discharge (10,000 tons)[1]	Total output value (10,000 yuan)[1]	Discharge intensity of output value (tons/10,000 yuan)[2]	Pollution control difficulty (yuan/ton)[3]	Wastewater pollution intensity (yuan/10,000 yuan)[4]
Recycling of waste resources and materials	8,381	1,457	2,069	2,624	0.79	5.75	4.54
Electricity and heat production and supply	248,143	367,984	158,928	47,353	3.36	0.67	2.26
Gas production and supply	4,469	1,248	989	3,142	0.31	3.58	1.13
Water production and supply	12,177	12,929	3,559	1,178	3.02	0.94	2.85

Notes 1. Refer to *China Environment Statistical Yearbook* (2012); 2. Discharge intensity of output value = Wastewater discharge/total output value; 3. Wastewater treatment difficulty = Operating costs of wastewater treatment facilities/wastewater treatment capacity; 4. Wastewater pollution intensity index = Emission intensity of output value × Wastewater treatment difficulty

Table 2.2 Industrial classification criteria

Discharge intensity of output value (ton/10,000 yuan)	>4.0	1.6–4.0	0.6–1.6	<0.6
Industrial category	High discharge	Relatively high discharge	Moderate discharge	Low discharge
Pollution control difficulty (yuan/ton)	>3.4	2.6–3.4	1.5–2.6	<1.5
Industrial type	High treatment costs	Relatively high treatment costs	Moderate treatment costs	Low treatment costs
Wastewater pollution intensity (yuan/10,000 yuan)	>9.0	3.0–9.0	1.3–3.0	<1.3
Industrial category	High pollution intensity	Relatively high pollution intensity	Moderate pollution intensity	Low pollution intensity

Note The table is reprinted from Huan et al. (2016), with permission from Research On Development

Table 2.3 Wastewater pollution types and characteristics by industrial sectors

Type	Sector	Characteristics
High pollution intensity	Textiles; beverage manufacturing; food manufacturing	High discharge, relatively high treatment costs
	Non-ferrous metal mining; chemical fiber manufacturing; chemical and chemical product manufacturing	High discharge, Moderate treatment costs
	Paper making and paper products; other mining	High discharge, Low treatment costs
	Pharmaceutical manufacturing	Relatively high discharge, relatively high treatment costs
	Textile apparel, footwear and hat manufacturing	Moderate discharge, high treatment costs
Relative high pollution intensity	Coal mining and washing	High discharge, Moderate treatment costs
	Leather, fur, feather (down) and their products	Relatively high discharge, relatively high treatment costs
	Oil processing, coking and nuclear fuel processing; rubber products	Relatively high discharge, Moderate treatment costs
	Agricultural and sideline food processing	Relatively high discharge, Low treatment costs
	Metal product manufacturing; recycling of waste resources and materials; manufacturing of communications equipment, computers and other electronic equipment	Moderate discharge, high treatment costs
	General equipment manufacturing	Low discharge, high treatment costs

(continued)

2.2 Water Pollution Characteristics of Different Sectors

Table 2.3 (continued)

Type	Sector	Characteristics
Moderate pollution intensity	Water production and supply; electricity and heat production and supply, non-metallic metal mining; ferrous metal mining	Relatively high discharge, Low treatment costs
	Cultural, educational and sporting goods manufacturing; transportation equipment manufacturing	Moderate discharge, relatively high treatment costs
	Oil and gas extraction; non-metallic mineral product manufacturing	Moderate discharge, Moderate treatment costs
	Printing and recording media replication	Low discharge, high treatment costs
Low pollution intensity	Ferrous metal smelting and calendering	Relatively high discharge, Low treatment costs
	Non-ferrous metal smelting and calendering	Moderate discharge, Low treatment costs
	Gas production and supply; electrical machinery and equipment manufacturing; furniture manufacturing	Low discharge, high treatment costs
	Wood processing and wood, bamboo, rattan, palm and grass products; manufacturing of instruments and cultural and office machinery; handicrafts and other manufacturing	Low discharge, relatively high treatment costs
	Special equipment manufacturing; tobacco products	Low discharge, Moderate treatment costs

Note The table is reprinted from (Huan et al. 2016), with permission from Research On Development

intensity was seen in the sectors of paper making and paper products, chemical and chemical product manufacturing, beverage manufacturing, leather, fur, feather (down) and their products, and chemical fiber manufacturing. It reached 2.07 and 1.52 tons per 100 million yuan in first two sectors respectively, far higher than others.

According to *China Environmental Statistical Yearbook* (2011), the largest emissions of heavy metals (lead, cadmium, mercury, total chromium, arsenic) came from non-ferrous metal smelting and calendering, leather, fur, feather (down) and their products, metal product manufacturing, and non-ferrous metal mining, which together accounted for 78.60% of the total in 2011 (China Environmental Statistical Yearbook 2011). Oil pollutant emissions were mostly sourced from coal mining and washing, ferrous metal smelting and calendering, oil processing, coking and nuclear fuel processing, chemical materials and chemical product manufacturing, as these sectors, which amounted to 57.2% of the total in 2011 (China Environmental Statistical Yearbook 2011).

References

Huan H, Li J, Li MX. Research on industrial structure adjustment of resource-based city from the perspective of groundwater pollution control: a case study in Jilin City [J]. Res Dev. 2016;6:106–12.

Ministry of Environmental Protection. Annual Report of Environmental Statistics (2011). http://www.zhb.gov.cn/gzfw_13107/hjtj/hjtjnb/201605/U020160604811616423937.pdf.

National Bureau of Statistics, Ministry of Environmental Protection. China environmental statistical yearbook 2012 [M].Beijing: China statistics press; 2016.

Chapter 3
Natural Circumstance and Industrial Economy of the Study Area

Abstract In order to determine and control the groundwater pollution risk from the perspective of industrial economy, the chapter expounds on the geographical location, hydro-meteorological conditions, soil and vegetation, terrain, landform, stratigraphy and structure, hydrogeological conditions and industrial characteristics of the study area.

Keywords Jilin city · Songhua river · Pore water · Precipitation
Water source areas · Heavy chemical industry · Low technical level
Significant spatial variation

3.1 Physical Geography

3.1.1 Geographical Location

The study area (43°52'24"–44°0'00"E and 126°26'5"–126°37'22.5"N) is located within the range of 5–10 km from the upstream Jilin Section of the Second Songhua River. It covers a total area of 104.5 km². On the right bank of the Songhua River is Longtan District of Jilin Province and on the left bank, Changyi District, as shown in Fig. 3.1.

3.1.2 Hydro-Meteorological Conditions

According to the meteorological data from 1981 to 2005 of Jilin Meteorological Station, the study area has an annual average temperature of 4.6 °C, annual average precipitation of 645.5 mm, and annual average evaporation of 1,506 mm. The annual maximum precipitation is up to 891.2 mm and minimum to 474.8 mm. The precipitation is uneven during the year, of which 70–80% takes place from June to September.

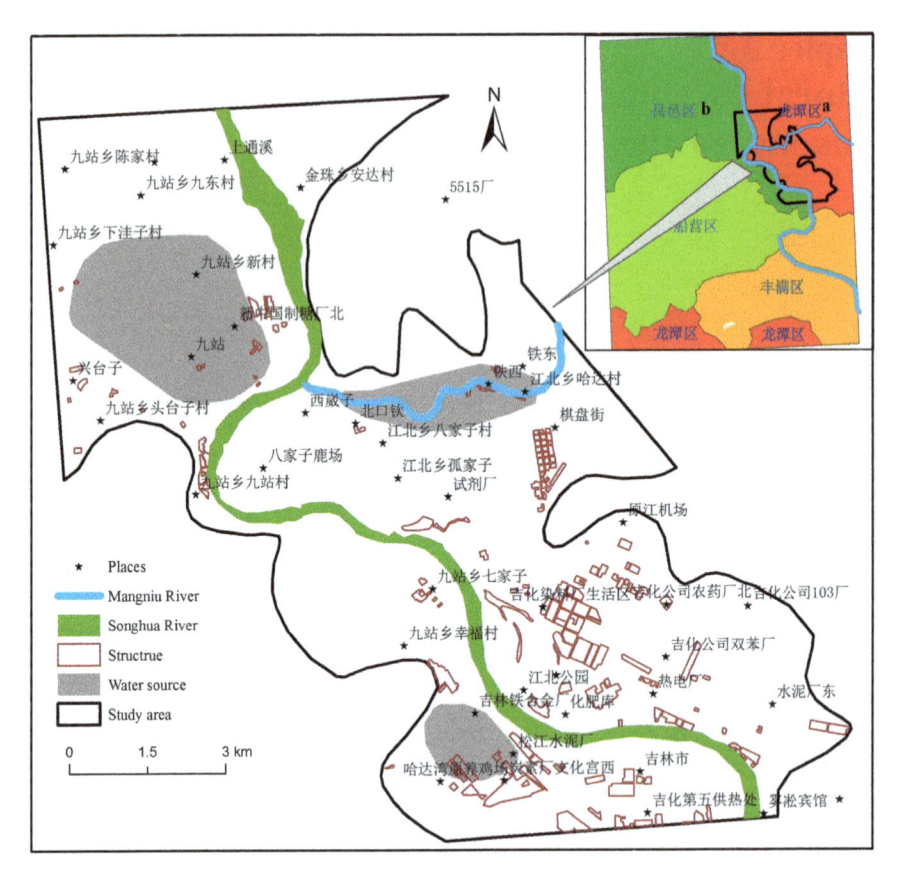

Fig. 3.1 Geographical location of the study area(*Note* **a** Longtan District; **b** Changyi District. The figure is reprinted from Huan et al. (2012), (2015), with permission from Elsevier and Springer respectively

River systems are advanced in the study area. Among them, the Songhua River flowing from southeast to northwest is the main stream and the Mangniu River is a large tributary. According to the 1979–2005 statistics of Jilin Hydrological Station, the Jilin section of the Songhua River has a stream gradient of 0.4%, annual average flow rate of 429.15 m^3/s, and annual average water level of 186.42 m. For the Mangniu River, the stream gradient reaches 2.7% and average flow rate 5.55 m^3/s.

3.1.3 Soil and Vegetation

The soils available in Jilin City are largely high-quality in a sound condition. They can be divided into five groups: dark brown soils, albic soils, alluvial soils, meadow soils and paddy soils. Forests, dominated by natural coniferous and broad-leaved forests, used to be distributed throughout the Longtan Mountain and surrounding hills in the eastern part of the study area. At present, there are basically no natural primitive forests and mostly natural secondary forests and plantations. Grassland is mainly scattered in a small scale along the Songhua River and its tributary Manguiu River, mostly in the form of floodplain. Farmland is widely distributed in the rural areas of rolling hills and river valley plains, covering dozens of crops, such as corn, rice, soybean and millet. Among them, rice is planted in the river valley plains and dryland farmed crops in the rolling hills.

3.2 Geological Conditions

3.2.1 Terrain and Landform

Overall, the terrain is high in the southeast and low in the northeast of the study area. The geomorphological characteristics are as shown in Table 3.1.

3.2.2 Stratigraphy and Structure

The strata of the study area are as shown in Table 3.2.

The study area is a part of the Jilin and Heilongjiang fold system of the Tianshan Hinggan geosynclinal region. There are dominantly north-west and north-east faults and east-west faults, and Yishu superimposed fault depression, Nanloushan Mesozoic superimposed fault depression and Mopanshan Mesozoic superimposed fault depression. These constructions of different properties, sizes and forms strictly

Table 3.1 Landforms

Cause	Form	Morphological unit	Geomorphological features and lithology
Stacked terrain	Alluvial plains	Second-order terrace	Distributed asymmetrically on both sides of the Songhua River and slightly tilted to the river bed. A binary structure with silt or silty clay in the upper part and gravel and cobble in the lower part. 185–190 m above sea level
		First-order terrace	Striped on both sides of the Songhua River and its tributaries. A binary structure, with silt or silty clay in the upper part and gravel and cobble in the lower part. 185 m above sea level
		Floodplain	Striped asymmetrically in the bed of the Songhua River and its tributaries, small-scale and composed of gravel and cobble. About 175 m above sea level

Table 3.2 Strata in key sections of the Songhua River

Age		Formation (strata) and reasons	Lithology
Quaternary	Holocene series (Q_4)	Alluvial strata (Q_4^{apl})	Distributed between the valleys of terraces and mountains; comprised of the upper pale yellow and black silt, and lower 0.5–20 cm-diameter pale yellow and variegated gravel and cobble; poorly separated, rounded and aqueous; sand and soil mixed; 4–12 m thick
		Alluvial strata ($Q_4^{1al,\ 2al}$)	Distributed widely and asymmetrically along the first-order terrace and floodplains of the Songhua River and Mangniu River valleys. The first-order terrace is comprised of 4–10 m-thick upper brown silt and 5–20 cm-diameter lower variegated gravel and cobble; 40 m thick to the most
	Upper Pleistocene series (Q_3)	Alluvial strata (Q_3^{al})	Mainly distributed in the second-order terrace of river valleys; typical binary structure of the 2–5 m-thick upper pale yellow clay and 5–10 cm-diameter lower sand and gravel; preferably separated and rounded; generally 5–15 m thick; good aquifer
	Middle Pleistocene series (Q_2)	Eluvial strata (Q_2^{dpl})	Distributed in the platforms inclined in the mountains, comprised of brown and golden-brown loess soil and silty clay; 5–20 m thick; vertical joint development, with local thin layer of fine sand lens; low water content
		Diluvial strata (Q_2^{apl})	Distributed in undulating platforms; binary structure that includes the upper exposed surface of 3–20 m brown and golden-brown loess clay, the thickness, lower gray silty clay and silt, and bottom 2–5 cm-diameter sand and gravel; round shape and poor separation; aqueous and 5–25 m thick
	Lower Pleistocene series (Q_1)	Diluvial strata (Q_1^{apl})	Discontinuously distributed in the downstream of Songhua River Valley and Mangniu River Valley and buried 10–20 m below the ground; comprised of variegated gravel, gravel, coarse sand, cobble and clay-containing rounded gravel; 2–12 m thick; good aquifer
		Fluvial strata (Q_1^{all})	Buried 10–40 m below the ground and distributed in Gudianzi, Jiuzhan and Jinzhu; controlled by Yishu trough structure and comprised of grayish and yellowish gravel sand, rounded gravel, cobble, soft sand and clay-containing rounded gravel; obvious phase transition; 10–30 m thick; mainly aquifers

(continued)

3.2 Geological Conditions

Table 3.2 (continued)

Age		Formation (strata) and reasons	Lithology
Upper triassic	Middle and upper eocene series	Manchurian ash formation (N_{1-2s})	Distributed in Gudianzi, Jiuzhan and Jinzhu and buried under the loose Quaternary strata 40–70 m below the ground; comprised of gray, greyish white loose weakly consolidated sand, sandstone and mudstone; more than 400 m thick and unconformity with the overlying lower Quaternary strata
Cretaceous	Lower series	Quantou formation (K_1q)	Buried below the loose stacked Quaternary strata in the northwest; comprised of purple and reddish brown mudstone, greyish white and slate-grey siltstone and sandstone; more than 400 m thick and unconformity with the upper Triassic Ash Formation
Permian	Upper permian	Yilaxi formation (P_2y_l)	Distributed in the Guoding Mountain and Huniuguo in the western city; comprised of neutral and acid volcanic rocks and interbedded tuff, sandstone and shale, intertwined with tuffaceous slate; 3,983 m thick and unconformity with the overlying Triassic Xiaofengmidingzi Formation strata
		Yangjiagou formation (P_2y)	Distributed in Fengman East and West Mountains, Shuangya Mountain, Suzaku Mountain, Langtou Mountain, Dongdawang, Yaolingzi and Laguta Ridge; mainly composed of siltstone, sandstone, and conglomerate and slate, and locally marble lens; 5,134 m thick and parallel unconformity with the lower and upper strata
	Lower permian	Fanjiatun formation (P_1f)	Distributed in Taipinggou and Yaogizi North–Xiaoha Hills and sporadic seen in Longtan Mountain, Taoyuan Mountain and Jianshan Mountain; mainly composed of soft metamorphic sandstone, siltstone, coarse gravel-tuffaceous sandstone, andesite clastic rocks, tuff and soft tuffaceous sandstone, and intertwined with thin-bedded limestone lens; 1,546 m thick and parallel unconformity with the overlying strata

control the formation and distribution of groundwater in this area. Affected by the tectonic movement, volcanic spills and activities are also very active in the area dated back to early Pleistocene of the Cenozoic Quaternary period. The sporadic distribution of early Pleistocene Xiaofengman olive, stomatal basalt, almond-shaped basalt on the eastern side of the Songhua River also reflects intense activities of the neotectonic movement.

3.2.3 Hydrogeological Conditions

3.2.3.1 Groundwater Endowment Characteristics

Groundwater in the study area can be divided into pore water from loose rocks and fissure and pore water from clastic rocks (Figs. 3.2 and 3.3), wherein the pore water from loose rocks supplies for industrial and agricultural production and domestic life. The natural conditions of groundwater formation determine the basic law of groundwater endowment and distribution.

(1) Pore water from loose rocks

The pore water from loose rocks is mainly distributed in the valley floodplain and first-order terrace of the Songhua River and the Mangniu River, existing in strongly permeable gravel, rounded gravel and cobble pores of Holocene, Upper and Lower Pleistocene series. Typically, the groundwater table is 4–7 m deep and

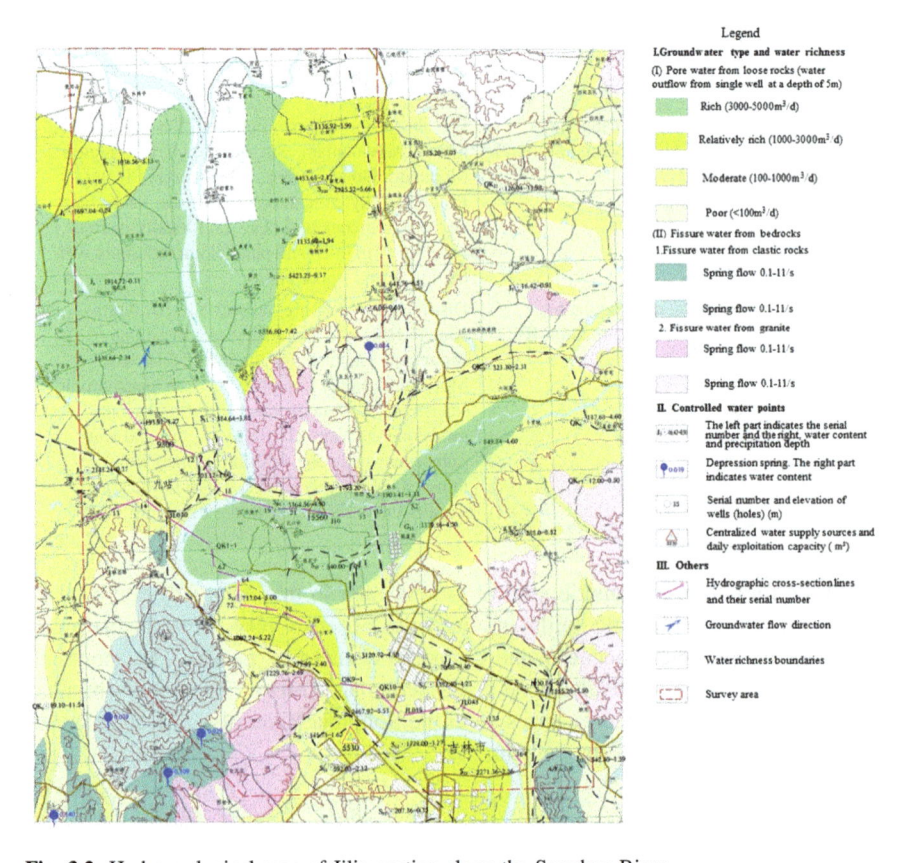

Fig. 3.2 Hydrogeological map of Jilin section along the Songhua River

3.2 Geological Conditions

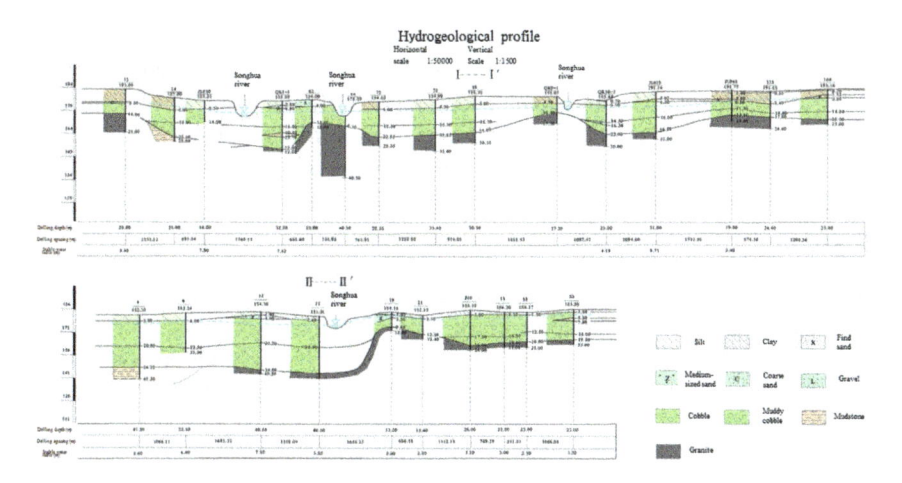

Fig. 3.3 Hydrogeological profile

the aqueous layer is 10–20 m thick, coarse and strongly conductive. The water outflow from single well reaches 3,000–5,000 m^3/d and even 5,000 m^3/d in certain plots when the depth is 5 m. While the river valley floodplain and first-order terrace are rich in water, the periphery floodplain and the leading edge of first-order terrace and second-order terrace has a moderate content of water (1,000–3,000 m^3/d). The groundwater table is 3–10 m deep and the aqueous layer of gravel and cobble, 10–15 m thick.

(2) Fissure and pore water from clastic rocks

The fissure and pore water comes from clastic rocks, including the Tertiary sandstone, conglomerate, siltstone distributed in Jiuzhan and Jinzhu. From southeast to northwest, the multilayer aquifer become thicker (10–30 m) and the top plate is buried deeper (25–50 m), with pressurized groundwater table of 5–10 m. Water richness is graded medium (100–500 m^3/d).

3.2.3.2 Groundwater Circulation Characteristics

Groundwater circulation in the area along the Songhua River is mainly controlled by meteorological, hydrological and topographical conditions, geological structure and human factors. Precipitation is the most important source of recharge. As the southeast is dominated by height-different moderate and low hills and the northwest by semi-closed mountain valley plains of the Songhua River and its tributaries, the groundwater flows from the southeast to the northwest and then beyond the area in the form of runoff. In the natural state, groundwater enters the river. The vertical recharge of pore water in loose rocks includes the infiltration of precipitation and irrigation water. The lateral recharge is derived from the groundwater runoff from waviness terraces and upstream valleys and river piracy of groundwater exploitation. In the natural state, the sound hydraulic conditions on both sides of the Songhua River and the Mangniu River and the permeable aquifer creates conditions for groundwater runoff, so groundwater flows to the river. In the area with intensive groundwater exploitation, the groundwater runoff accelerates and changes the groundwater flow, and the river replenishes the groundwater. Due to poor permeability of the loess-like silty clay layer in the terrace, the conditions for groundwater runoff are far from desired.

Evaporation is weak in the area as groundwater is mined in Jiuzhan, Hadawa and Songyuan Hada while recharging surface water drainage. According to existing data, the maximum depth for phreatic evaporation is 4.95 m, so groundwater discharge by means of evaporation mainly occurs in the rear edge of terraces north of Zunyi Road.

3.2.3.3 Groundwater Dynamic Characteristics

(1) Intra-annual variance of groundwater table

The study of groundwater dynamic characteristics focuses on pore groundwater from Quaternary loose rocks. According to the factors of pore groundwater circulation, groundwater is grouped into five types (Table 3.3), of which the first four occur in non-concentrated exploitation areas with a low level of exploitation and the latter in areas with a high level of concentrated exploitation (Jiuzhan Industrial Zone, Hadawan Industrial Zone and Songyuan Hada water source area).

(2) Inter-annual variance of groundwater table

In the non-centralized exploitation area, the inter-annual change of groundwater table is subject to precipitation. More precipitation implies less depth to the water table, and vice versa. In the concentrated exploitation area, the inter-annual change is mainly controlled by exploitation. Exploitation decreases the water table, and vice versa.

In general, there is no downward trend in the groundwater table over the years in Jilin City regardless of concentrated exploitation, though rises and falls occur in terms of intra-annual variation.

3.3 Groundwater Exploitation and Utilization

Jilin City relies on the Songhua River for water supply, and enjoys very favorable conditions for riverside water intake owning to very rich groundwater resources. In 2015, the water consumption of its urban area amounted to 12.41×10^8 m^3/a, of which 1.722×10^8 m^3/a or 14.3% was sourced from groundwater. Currently, the water supply capacity reaches 18.72×10^8 m^3/a, of which groundwater took up

3.3 Groundwater Exploitation and Utilization

Table 3.3 Types and characteristics of pore groundwater movement

Type	Distribution	Recharge	Discharge	Groundwater table intra-variation
Infiltration–evaporation	Rear edge of the first and second-order terraces of the Songhua River in Northeast Jiangbei Chemical Industrial Zone and Jiangbei Township, and terraces of the Mangniu River	Precipitation and farmland recharge	Evaporation	Low level (in February-June) due to very little precipitation and intense evaporation from October to the next May; high level in July-September under the impact of post-June precipitation
Infiltration–runoff (most common)	Jiuzhan and Jiangbei Chemical Industrial Zones and first and second-order terraces of the Songhua River	Precipitation and farmland recharge	Lateral runoff	Later peak and smaller intra-annual change than infiltration-evaporation groundwater; low level in January-June
Infiltration–Intermittent exploitation	sprinkling area of vegetable irrigated area and industrial exploitation in both sides of Songhua River	Precipitation	Artificial intermittent exploitation	A peak and several troughs in a year. High level in non-irrigation, late rainy season, with peak in July-August; low level in February-May with intermittent exploitation
Runoff infiltration–runoff	Undulating terraces on both sides of the Songhua River and its tributaries	Precipitation and lateral runoff in low hilly areas	Runoff	A peak and a trough in a year. High level in the late rainy season with peak in July-December, and low level in April-June
Lateral river seepage–exploitation	Haidawan Industrial Zone and Jiuzhan Industrial Zone on the west bank of the Songhua River, and the Songyuan Hada water source area on the north bank of the Mangmiu River	Lateral runoff with the Songhua River and the Mangniu River as the fixed water head boundary for recharge	Artificial exploitation	Undulating fluctuations synchronized with river level changes

2.12×10^8 m^3/a or 11.3%. Groundwater serves as an important source of water supply for livestock and poultry in rural areas and towns, inrrigation of farmland and paddy field, areas without access to inadequate municipal water supply, and areas with contaminated surface water bodies.

As far as the study area is concerned, groundwater exploitation mainly takes place in such water source areas as Songyuan Hada, Hadawan and Jiuzhan Industrial Zone. Table 3.4 shows the picture of groundwater exploitation in 2005.

Table 3.4 Groundwater exploitation in industrial zones in 2005

Item Water source		Volume of exploitation ($10^4 m^3/a$)	Recharge resources ($10^4 m^3/a$)	Utilization rate (%)	Time of production
Mangniu River	Songyuan Hada Industrial Zone	567.94	1,426.68	39.81	1992
Songhua River Basin	Hadawan Industrial Zone	201.85	635.83	31.75	1964
	Jiuzhan Industrial Zone	357.70	991.63	37.81	1953

Groundwater is exploited for industrial purpose in Songyuan Hada and Hadawan and for both industrial and domestic purposes in Jiuzhan. There is still a large potential for groundwater exploitation in the three places.

3.4 Industrial Characteristics of Jilin City

3.4.1 Heavy Reliance on Resources

Abundant mineral resources in Jilin City and the surrounding areas support the important role of resource-based advantageous industries in regional economic development. There are 50 varieties of mineral resources identified in the city, 14 large mineral deposits, 27 medium-sized mineral deposits, and more than 1,000 mineral spots. Among them, 26 varieties are of very importance for the city and 9 for the county, including iron, molybdenum, nickel, gold, limestone, wollastonite, decorative granite, refractory clay, and aphanitic graphite. Energy minerals are most distributed in the northern Wanchang–Shulan area, non-ferrous minerals in central Daheishan and Hongqiling area, ferrous metals and precious metals in the southern Jiapigou area of Huadian City, and non-metallic minerals in rock-based areas. In 2005, the output of raw coal in Jilin City attained 2.268 million tons, pig iron 1.084 million tons and gold 319,000 tons. Resource extraction industries produced 1.567 billion yuan, accounting for 2.08% of industrial output in the region.

Rich natural resources create favorable conditions for the development of the chemical industry in Jilin City. Based on coal and hydropower resources and nearby oil and gas resources, the city has developed a petrochemical–chemical fiber–textile and garment industrial chain and established the country's important ethylene industrial base, chemical fiber base and fine chemical base. Relying on the local iron ore, molybdenum and nickel resources, the city also expands the ferrous and non-ferrous metal smelting sector to a considerable scale and secures an important position in fields of metallurgical furnace charge, carbonyl metals, and fine steel. The industries of non-metallic mineral mining and products are relatively developed owning to the huge reserves. Jinlin City has an advantage in the reserves of wollastonite, limestone, aphanitic graphite, refractory clay, calcite, decorative granite. Among them, wollastonite is mainly exported to Western Europe and the United States and calcite to South Korea and Japan and refractory clay and "Jilin white marble" have a high profile on the international stone market. In addition, the country's important grain production base embraces rich biodiversity and obvious advantage in natural wild resources. Based on this, Jilin City has developed a large and diverse food and beverage industry and formed significant processing and conversion capacity in fields of beer, dairy, wine, and fuel ethanol.

3.4.2 Clear Dominance of Heavy Chemical Industry

Jilin is an industrial city built during the 1st and 2nd FYP periods. It accommodated 7 of the 156 key projects in the country, including the construction of large industrial enterprises, such as dye plant, fertilizer plant, calcium carbide plant, and ferroalloy plant, carbon plant. This has paved the foundation for the development of heavy industry, dominated by chemical, metallurgy, and electric power industries. In 2005, the basic industrial framework took shape after a long period of accumulation and construction. It encompasses the four pillars (chemical, automotive, metallurgy, food industries), four advantageous industries (medicine, building materials, textile and energy), and two emerging industries (electronic information and new materials). The development of specific industries is described in Table 3.5.

The industrial structure of Jilin City is dominated by heavy chemical industry. The most productive eight sectors account for 88.13% of the total industrial output value. They are chemical and chemical product manufacturing, ferrous metal smelting and calendering, chemical fiber manufacturing, electricity and heat production and supply, transportation equipment manufacturing, non-metallic mineral products, non-ferrous metal smelting and calendering, and beverage manufacturing. Among them, the chemical industry is the largest leading industry in the city. There have been business clusters underpinned by PetroChina Jilin, JCGC, Jilin Fuel Ethanol Co., Ltd, and Jilin Chemical Fiber Group Co., Ltd. The industry is capable refining 7 million tons of oil per year, producing 530,000 tons of ethylene, 300,000 tons of fuel ethanol and 263,000 tons of viscose staple fiber, viscose filament yarn, acrylic fiber and pulp fiber.

Ferrous metal smelting is the second largest pillar industry in Jilin City. It develops rapidly, driven by the expansion of Jilin Ferroalloy Co., Ltd. and Jianlong Iron and Steel Co., Ltd. and the investment of Jilin Huaqi Pipe Co., Ltd. The annual steel production capacity reaches 1.2777 million tons and ig iron, 1.0842 million tons. In addition, with adequate hydropower and coal power, the total installed capacity attains 4 million kilowatts in the city, involving Longtan Thermal Power Plant, Bashan Power Plant, Fengman Power Plant. Relying on FAW Group, the automobile industry also reaches a certain scale, including FAW Jilin Light Vehicle Depot. It has a production capacity of 90,000 complete vehicles a year and

Table 3.5 Development of major industries in Jilin City

Industry	Description
Chemical industry	Development from the initial dye plant, fertilizer plant, and calcium carbide plant to comprehensive petrochemical companies, of which PetroChina Jilin and Jilin Chemical Group Corporation (JCGC) lead the rapid development of the whole industry
Metallurgical industry	China's largest ferroalloy production enterprise and second largest nickel metal production base, with the establishment of Jianlong Iron and Steel Co., Ltd., and Sinosteel Corporation's acquisition of Jilin Ferroalloy Co., Ltd.
Chemical fiber industry	A comprehensive production capacity up to 446,000 tons, represented by Chemical Fiber Group, Leading products involve over 450 varieties in 6 series, including viscose staple fiber, viscose filament yarn, acrylic fiber, pulp fiber, yarn, and paper products, and are exported to over 10 countries and regions, such as Japan, Korea, Europe and the United States
Automobile industry	An important production base of light cars and family cars, represented by FAW Jilin Light Vehicle Depot. The capacity covers five series of products, including light vehicles, minibuses, economical cars, special vehicles, and agricultural vehicles, and six series of ancillary products, including chassis, interior parts, electrical parts, non-metallic parts, engines and gearboxes, and lights
Pharmaceutical industry	A base for modernization of Chinese medicine, and an important place of origin for chemicals and biological drugs. The production capacity covers chemicals, Chinese and Western medicine preparations, medical equipment, medical packaging and health materials
Food industry	Initial industrial formation with such backbones as alcohol, beer, cold drinks, ice cream, rice, and edible fungi, and emergence of new sectors of processing Chinese forest frogs, wild vegetables, and mineral water based on corn, soybean and rice processing

increases the output value of auto parts enterprises to 1 billion yuan. The output value of related enterprises is as shown in Table 3.6.

By comparing the industrial structure of Jilin City and the country in 2005 (Table 3.7), the study finds industries with high pollution intensity take up a significantly large proportion (66.48%), 45.40% higher than the national average (21.08%). As to industries with moderate and low pollution intensity, the respective proportion is small and adds up to 29.11% only, 16.61% lower than the national average (45.73%). In terms of sectors (Table 3.8), the large percentage of high-pollution-intensity industries can be attributed to the relatively high proportion of developed chemical industry, especially manufacturing of chemical materials, chemical products and chemical fiber. In 2005, the chemical and chemical product manufacturing contributed 57.27% of the total industrial output value, 8.8 times the national average, and became the absolutely dominant industrial sector in the region. Only second to this sector was chemical fiber manufacturing, which accounted for 4.81%. They were followed by ferrous metal smelting and calendering and electricity and heat production and supply.

3.4 Industrial Characteristics of Jilin City

Table 3.6 Jilin-based industrial enterprises with largest output value in 2005

Industrial enterprises	Output value (10,000 yuan)	Industrial enterprises	Output value (10,000 yuan)
PetroChina Jilin	3,249,606	Daheishan Molybdenum Co., Ltd of Jilin Nickel Industry Group	20,142
JCGC	650,279	Longda Ferroalloy Co., Ltd of Jilin Province	19,193
Jilin Chemical Fiber Group Co., Ltd	363,278	Fukang Wood Industry Co., Ltd of Jilin City	19,171
Jilin Jianlong Iron and Steel Co., Ltd	331,587	Jilin Tuopai Agricultural Product Development Co., Ltd	19,065
Jilin Ferroalloy Co., Ltd	226,196	Jilin Far East Pharmaceutical Group	18,977
FAW Jilin Light Vehicle Depot	150,135	JCGC Cement Plant	17,230
Jilin Carbon Group Co., Ltd	132,531	Shulan Mining Bureau	16,719
Longtan Thermal Power Plant of Jilin Longhua Thermal Power Co., Ltd	125,233	Jiaoligou Gold Mining Co., Ltd	15,828
Jilin Jien Nickel Industry Co., Ltd	124,000	Jilin Longding Electric Co., Ltd.	15,323
Jilin Fuel Ethanol Co., Ltd.	118,065	Baishishan Forestry Bureau of Jilin Province	15,280
Panshi Seamless Steel Tube Co., Ltd of Tonghua Iron and Steel Group	73,000	Shulan Synthetic Pharmaceutical Co., Ltd of Jilin Province	14,432
Jilin Jingwei Steel Manufacturing Co., Ltd	60,171	JCGC Jilin Longshan Chemical Plant	14,431
Jilin Huaxing Electronics Group Co., Ltd	58,191	Jilin Municipal Thermal Corporation	14,257
Baishan Power Plant of Huadian City	56,251	Jilin Yongda Group Co., Ltd	14,200
Jilin Jiangbei Manufacturing Co., Ltd	42,142	Jilin Kelong Electric Appliance Co., Ltd	14,039
Jilin Huaqi Tube Co., Ltd	38,281	Jilin Longshan Organic Silicone Co., Ltd.	13,836
Fengman Power Plant	37,181	Jilin Hydraulic Machinery Co., Ltd	12,230
JCGC Jinjiang Oil Plant	36,672	Jilin Pharmaceutical Co., Ltd.	10,020
Jilin Thermal Power Plant	33,465	Huadian Mining Co., Ltd of Tonghua Iron and Steel Group	9,712
Jilin Yatai Mingcheng Cement Co., Ltd	33,252	Jilin Municipal Water Co., Ltd	9,359
China Resources Snow Breweries (Jilin) Co., Ltd.	31,483	Jilin Cement Sleeper Factory of Shenyang Railway Bureau	9,318
Jidong Cement (Jilin) Co., Ltd	29,514	Jilin Xinye Equipment Co., Ltd	9,272
Huadian Jianlong Mining Co., Ltd	26,814	Jilin Jiuniu Dairy Co., Ltd	8,310

(continued)

Table 3.6 (continued)

Industrial enterprises	Output value (10,000 yuan)	Industrial enterprises	Output value (10,000 yuan)
Hongshi Forest Branch of Jilin Forest Industry Co., Ltd	23,871	Jilin Dongguan Thermal Power Plant	7,797
No. 5704 Factory of China Aviation Industry Corporation I	20,915	Shulan Crane Accessories Co., Ltd	6,805

Source Social and Economic Statistical Yearbook of Jilin City (2006)

Table 3.7 Industrial structure comparison between Jilin City and the country in 2005

Category	Jilin City[1]		Country[2]	
	Total output value (10,000 yuan)	Percentage (%)	Total output value (100 million yuan)	Percentage (%)
High pollution intensity	5,017,209	66.48	53,043.77	21.08
Relatively high pollution intensity	314,414	4.17	78,452.2	31.18
Moderate pollution intensity	1,034,083	13.70	54,232.84	21.55
Low pollution intensity	1,163,252	15.41	60,822.81	24.17

Source 1. Calculations based on *Social and Economic Statistical Yearbook of Jilin City* (2006); 2. Calculations based on *China Statistical Yearbook* (2006)

3.4.3 Relatively Low Technical Level

Raw material industries take a major part of industrial development in Jilin City. Many products have low technological content and added value, and the extensive growth model with over-reliance on resources and capital investment is very common. In 2004, only 47 enterprises were found to carry out scientific and technological research and development, accounting for 9% of the surveyed total; and only 4% of the full-time staff participated in science and technology activities. The total investment in scientific and technological research and development numbered 400 million yuan, less than 1% of the regional gross domestic product. The technical level of industrial development can be measured by total factor productivity (TFP) which reflects the increase in the value created by capital and labor inputs. The total factor productivity is calculated as follows: industrial added value/[(average annual balance of floating assets + annual average balance of net fixed assets) × 0.5 × (number of employees) × 0.5].

According to the TFP comparison of Jilin City and the country's major industrial sectors in 2015 (Table 3.9), most of the industries (78.8%) had a technical level below the national average, apart from metal product manufacturing, plastic product

3.4 Industrial Characteristics of Jilin City

Table 3.8 Comparison of high-pollution-intensity industrial sectors in Jilin City and the country in 2005

Industrial sectors	Jilin City[1]		Country[2]	
	Total output value (10,000 yuan)	Percentage (%)	Total output value (100 million yuan)	Percentage (%)
Non-ferrous metal mining	49,540	0.66	1,140.41	0.45
Food manufacturing	55,391	0.73	3,779.39	1.50
Beverage manufacturing	112,911	1.50	3,089.27	1.23
Textiles	17,423	0.23	12,671.65	5.04
Garment, shoe, hat manufacturing	2,500	0.03	4,974.63	1.98
Paper making and paper product manufacturing	13,656	0.18	4,161.33	1.65
Chemical and chemical product manufacturing	4,322,307	57.27	16,359.66	6.50
Pharmaceutical manufacturing	80,203	1.06	4,250.45	1.69
Chemical fiber manufacturing	363,278	4.81	2,608.39	1.04

Source 1. Calculations based on *Social and Economic Statistical Yearbook of Jilin City* (2006); 2. Calculations based on *China Statistical Yearbook* (2006)

manufacturing, ferrous metal mining, and oil processing, coking and nuclear fuel processing. The TFP value is as low as 0.73, 0.83, 0.84, 0.62, and 0.57 of the national average in chemical and chemical product, ferrous metal smelting and calendering, electricity and heat production and supply, transportation equipment manufacturing and non-metallic mineral product manufacturing.

3.4.4 Significant Spatial Variation of Industry

Table 3.10 illustrates the industrial distribution in some areas of Jilin City in 2005. There were most heavy industry enterprises in Longtan District, Fengman District and Chuanying District, numbering 33, 24 and 21 respectively. According to the number of enterprises, the ratio of light and heavy industries was significantly higher in the Economic and Technological Development Zone, High-Tech Zone, and Longtan District, reading 14.00, 6.33 and 4.13 respectively. Due to large economic proportion and many heavy industry and chemical enterprises, Longtan District produced the 76.57 million tons of wastewater in 2005, second only to Changyi District (77.45 million tons), and 1,179 tons of ammonia nitrogen, at the same level with the other three districts (Longtan District, Fengman District, Chuanying District). Changyi District discharged the most wastewater of up to 45 million tons, of which COD reached 18,433 tons, far more than others, and

Table 3.9 Total factor productivity comparison of Jilin City and the country's major industrial sectors in 2015

Industry	Jilin [1]	Country [2]	Ratio
Metal product manufacturing	3.58	1.78	2.01
Plastic product manufacturing	2.52	1.54	1.64
Ferrous metal mining	3.77	2.40	1.57
Oil processing, coking and nuclear fuel processing	4.84	3.16	1.53
Rubber product manufacturing	2.18	1.65	1.32
Non-ferrous metal smelting and calendering	2.60	2.32	1.12
Handicrafts and other manufacturing	1.60	1.50	1.07
Chemical fiber manufacturing	1.64	1.65	1.00
General equipment manufacturing	1.60	1.73	0.92
Water production and supply	0.74	0.82	0.90
Non-ferrous metal mining	2.25	2.51	0.90
Food manufacturing	1.78	2.07	0.86
Electricity and heat production and supply	1.79	2.12	0.84
Ferrous metal smelting and calendering	2.35	2.84	0.83
Agricultural and sideline food processing	2.16	2.66	0.81
Leather, fur, feather (down) and their products	1.18	1.53	0.77
Beverage manufacturing	1.76	2.30	0.76
Chemical and chemical product manufacturing	1.57	2.16	0.73
Manufacturing of communications equipment, computers and other electronic equipment	1.55	2.20	0.70
Furniture manufacturing	1.07	1.55	0.69
Textiles	0.93	1.43	0.65
Special equipment manufacturing	0.98	1.57	0.62
Transportation equipment manufacturing	1.12	1.81	0.62
Non-metallic mineral product manufacturing	0.85	1.48	0.57
Wood processing and wood, bamboo, rattan, palm and grass products	0.94	1.67	0.56
Manufacturing of instruments and cultural and office machinery	0.96	1.81	0.53
Pharmaceutical manufacturing	1.12	2.12	0.53
Non-metallic mining	0.94	1.86	0.51
Paper and paper products	0.70	1.61	0.43
Electrical machinery and equipment manufacturing	0.82	1.96	0.42
Textile apparel, footwear and hat manufacturing	0.53	1.47	0.36
Coal mining and washing	0.54	1.72	0.31
Oil and gas extraction	1.60	6.92	0.23

Source 1. Calculations based on *Social and Economic Statistical Yearbook of Jilin City* (2006); 2. Calculations based on *China Statistical Yearbook* (2006)

3.4 Industrial Characteristics of Jilin City

Table 3.10 Industrial distribution comparison between districts of Jilin City in 2005

Administrative division	Enterprises above the designated scale	Light and heavy industries		Heavy/ light industry
		Light industry	Heavy industry	
Changyi District	27	17	10	0.59
Longtan District	41	8	33	4.13
Chuanying District	32	11	21	1.91
Fengman District	35	11	24	2.18
High-tech Zone	22	3	19	6.33
Economic and Technological Development Zone	15	1	14	14.00

Source Calculations based on *Social and Economic Statistical Yearbook of Jilin City* (2006)

ammonia nitrogen 990 tons, lower than Longtan District. Chuanying District registered the least wastewater discharge of 7.47 million tons, while Fengman District 20.63 million tons, higher than Chuanying, but lower than Changyi District and Longtan District.

In 2005, the output value of enterprises above the designated scale in Longtan District attained 2.372 billion yuan, mainly driven by chemical enterprises and auto parts enterprises. As one of the country's largest production base of chemical raw materials, Longtan Chemical Park has made remarkable achievements with increasing effect of industrial agglomeration. It accommodates more than 40 large and medium-sized enterprises, such as PetroChina Jilin, Jilin Thermal Power Plant with an installed capacity of 900 MW, state-run Jiangbei Machinery Plant, and China Resources Breweries Group. In Fengman District, enterprises above the designated scale created an output value of 2.316 billion yuan, mainly distributed in the sectors of agricultural processing, auto parts manufacturing, wood processing, and building materials. Changyi District is an old town and the business center and transportation hub of Jilin City. It witnessed an output value of 929 million yuan of enterprises above the designated scale in 2005. Private industrial enterprises form clusters and the light industry takes up higher proportion. The representative enterprises in the district include Jilin Ferroalloy Co., Ltd and Jilin Carbon Co., Ltd. Chuanying District is the birthplace and the site of the ancient city of Jilin and has a developed service industry. In 2005, the output value of enterprises above the designated scale reached 756 million yuan. In addition to prominent light industry, the advantageous industries include machinery processing, food and medicine, building materials and decoration, while the main products are high voltage switches, dairy products, wine. In 2005, the High-Tech Zone achieved an industrial output value of 50 billion yuan and an investment of of 4.18 billion yuan in fixed assets, of which 2.32 billion yuan has been in place. It has fostered four industrial pillars, namely fine chemicals, motor vehicles and parts manufacturing, electronic information, and biological medicine, and built a large-scale chemical production base dominated by coal chemical industry. The new projects include the 1-million-tons/year delayed coking project, 70,000-tons/year aniline project and 600,000-tons/year ethylene expansion project. In 2015, the Economic and Technological Development Zone, underpinned by synthetic textiles, biotechnology and fine chemicals, yielded an industrial output value of 7.6 billion yuan and leveraged domestic investment of 3.04 billion yuan and foreign investment of 20.5 million US dollars.

References

Editorial Board of Social and Economic Statistical Yearbook of Jilin City. Social and economic statistical yearbook of Jilin City [M]. Beijing: China Statistics Press, 2006.

Huan H, Wang JS, Lai DS, et al. Assessment of well vulnerability for groundwater source protection based on a solute transport model: a case study from Jilin City, northeast China[J]. Hydrogeol J. 2015;23:581–96.

Huan H, Wang JS, Teng YG. Assessment and validation of groundwater vulnerability to nitrate based on a modified DRASTIC model: A case study in Jilin City of northeast China[J]. Sci Total Environ. 2012;440:14–23.

National Bureau of Statistics. China statistical yearbook [M]. Beijing: China Statistics Press; 2006.

Chapter 4
Groundwater Pollution Characteristics and Source Apportionment

Abstract The chapter elaborates on the analysis of groundwater hydro-chemical characteristic, assessing the groundwater quality and groundwater pollution according to the groundwater quality data collected in the 2005 and judging the sources of groundwater pollution according to the distribution of land use, pollutant sources and concentration of typical components which exceed the groundwater quality standard. The results will guide the following validation of groundwater pollution risk mapping and making the appropriate risk control measures.

Keywords Nemerow's index · Inorganic component · Organic component Superposition grading index method · Groundwater pollution sources

4.1 Hydro-Chemical Characteristics

Groundwater hydro-chemical characteristic analysis and groundwater quality and pollution evaluation are conducted with 24 groundwater samples collected in the normal period in 2005 (Fig. 4.1). The analysis covers 34 inorganic components and 26 organic components, including 26 volatile organic compounds (VOCs), 11 organochlorine pesticides, and benzo [a] pyrene (BaP) of polycyclic aromatic hydrocarbons (PAH). All the data are sourced from Jilin Geological Environmental Monitoring Station. The results of inorganic and organic component analysis are as shown in Tables 4.1 and 4.2.

Table 4.1 indicates that the groundwater environment ranges from weak acid to weak base. Hg, NO_2–N, NH_4–N, and Cr^{6+} have the highest coefficient of variation over 200%, and their distribution is uneven and subject to man-made sources. The average concentrations of NO_3–N, NH_4–N, TFe and Mn exceed Class III groundwater quality standards by 2.94, 6.24, 2.88 and 3 times respectively and the maximum concentrations by 11.29, 62.22, 23.9 and 16.3 times respectively. Since the background values in the study area exceed the standards, the TFe and Mn concentrations beyond the standards are affected by the native hydrogeological environment. Among inorganic components, SO_4^{2-} and Hg show most-fold

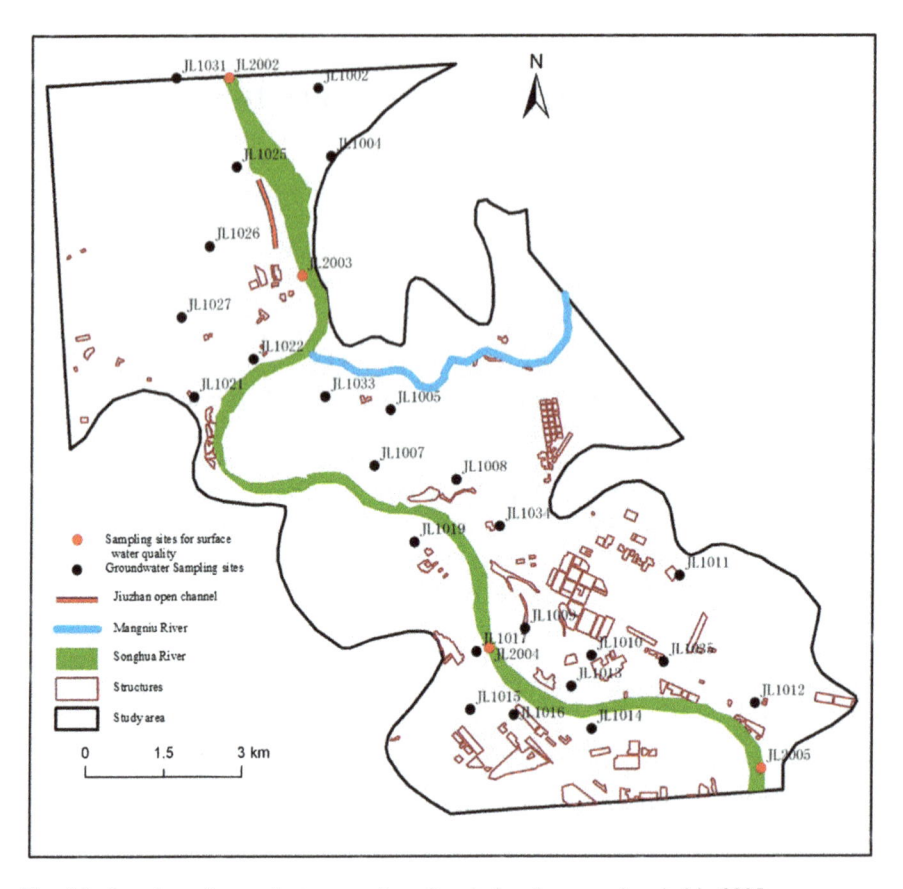

Fig. 4.1 Location of groundwater sampling sites during the normal period in 2005

maximum concentrations, up to 157.25 and 340 times of the background values respectively, followed by NO_3–N (73.29 times), Cl^- (20.30 times) and Cr^{6+} (17.72 times) and total dissolved solids (TDS, 14.04 times). It implies that these components are largely influenced by human activities.

As shown in Table 4.2, the organic components detected in descending order of times are chloroform (11), BaP (9), trichlorethylene (6), chlorobenzene (5), total BHC (4), and other components (3 or less). Among them, chlorobenzene has the highest coefficient of variation in concentration (79.87%), successively followed by vinyl chloride (16.26%) and trichlorethylene (15.43%), and other organic components (less than 5%). The average concentrations of carbon tetrachloride, benzene and BaP exceed the Class III groundwater quality standards by 1.62, 2.88 and 4.54 times respectively, and the maximum concentrations by 2.76, 4.97, and 31.40 times. The maximum concentrations over 100 times than background values are found, in descending order, in chlorobenzene, ethylbenzene, vinyl chloride, benzo[a] pyrene, trichlorethylene, benzene, and 1, 2-dichloroethylene.

4.1 Hydro-Chemical Characteristics

Table 4.1 Statistical characteristics of inorganic components of groundwater in 2005 (n = 24)

Indicator	Minimum value (mg/L)	Maximum value (mg/L)	Average value (mg/L)	Standard deviation (mg/L)	Variation coefficient (%)	Number of detection	Class III standards (mg/L)	Number of wells exceeding the standards	Non-compliance rate (%)	Maximum times of exceeding the standards	Background value (mg/L)	Number of wells exceeding the standards	Non-compliance rate (%)	Maximum times of exceeding the standards
pH	6.71	8.04	7.21	0.37	5.17	24	6.5–8.5	0	0.00		7.06	14	58.33	1.14
Total hardness (TH)	80.19	657.54	342.34	167.65	48.97	24	450	7	29.17	1.46	159.84	20	83.33	4.11
TDS	318.37	1,470.01	702.06	307.69	43.83	24	1,000	3	12.50	1.47	104.712	24	100.00	14.04
SO_4^{2-}	3.85	454.3	92.48	98.9	106.95	23	250	2	8.33	1.82	2.889	24	100.00	157.25
Cl^-	7.19	251.64	76.23	56.67	74.34	24	250	0	0.00		12.399	23	95.83	20.30
NO_3-N	0.677	112.9	29.41	28.98	98.52	24	10	16	66.67	11.29	1.53	23	95.83	73.79
NO_2-N	0.0003	0.03	0.0034	0.0072	212.85	22	0.2	0	0.00			24	100.00	
NH_4-N	0.0031	31.112	3.12	8.84	283.29	12	0.5	5	20.83	62.22		24	100.00	
COD	0.77	4.26	1.49	0.79	52.74	24	3	1	4.17	1.42	1.267	12	50.00	3.36
I	0.0007	0.0086	0.0052	0.0028	52.50	12	1	0	0.00		0.003726	23	95.83	2.31
F	0.18	0.62	0.31	0.14	45.71	24	1	0	0.00		0.354	6	25.00	1.75
Hg	0.0001	0.0068	0.0008	0.0016	208.94	22	0.001	4	16.67	6.80	0.00002	24	100.00	340.00
As	0.0001	0.0075	0.0014	0.0019	135.61	24	0.01	0	0.00		0.000788	11	45.83	9.52
Cr^{6+}	0.0003	0.0118	0.0058	0.017	291.92	23	0.05	0	0.00		0.000666	20	83.33	17.72
Cd	0.0002	0.0003	0.0002	2E-05	9.99	24	0.005	0	0.00		0.000046	24	100.00	6.52
Pb	0.0001	0.0031	0.0005	0.0007	135.47	24	0.01	0	0.00		0.000338	6	25.00	9.17
Zn	0.0054	0.0648	0.015	0.015	98.00	24	1	0	0.00		0.012653	7	29.17	5.12
Al	0.0004	0.831	0.078	0.18	234.09	24	0.2	3	12.50	4.16		24	100.00	
TFe	0.03	11.95	1.43	3.18	221.92	24	0.5	9	37.50	23.90	3.674	2	8.33	3.25
Mn	0.01	4.89	1.00	1.47	146.81	17	0.3	8	33.33	16.30	0.98	5	20.83	4.99

Table 4.2 Statistical characteristics of organic components of groundwater in 2005 (n = 24)

Index	Descriptive statistics					Comparison with Class III groundwater quality standards					Comparison with limits of detection		
	Minimum value (mg/L)	Maximum value (mg/L)	Average value (mg/L)	Standard deviation (mg/L)	Variation coefficient (%)	Number of detection	Class III standards (mg/L)	Number of wells exceeding the standards	Non-compliance rate (%)	Maximum times of exceeding the standards	Limit of detection (mg/L)	Number of wells exceeding the standards	Non-compliance rate (%)	Maximum times of exceeding the standards
Vinyl chloride	0.00147	0.00167	0.00157	0.0137	16.26413	2	0.005	0	0		0.0005	2	8.33	3.34
1,1-dichloroethylene	0.00021	0.00032	0.0003	6E-09	0.002283	2	0.03	0	0		0.0002	2	8.33	1.60
1,2-dichloroethylene	0.00171	0.0265	0.0141	0.0003	2.178462	2	0.05	0	0		0.0002	2	8.33	132.50
Chloroform	0.00021	0.00574	0.0019	3E-05	1.405181	11	0.06	0	0		0.0002	11	45.83	28.70
Carbon tetrachloride	0.00097	0.00551	0.0032	1E-05	0.31808	2	0.002	1	4.17	2.76	0.0002	2	8.33	27.55
1,2-dichloroethane	0.00186	0.0116	0.0069	5E-05	0.689265	3	0.03	0	0		0.0003	3	12.50	38.67
Benzene	0.0079	0.0497	0.0288	0.0009	3.033403	2	0.01	1	4.17	4.97	0.0003	2	8.33	165.67
Trichloroethylene	0.0017	0.0564	0.0144	0.0022	15.42625	6	0.07	0	0		0.0002	6	25.00	282.00
1,2-dichloropropane	0.0004	0.0004	0.0004	0	0	1	0.005	0	0		0.0002	1	4.17	2.00
Toluene	0.00094	0.0034	0.0022	3E-06	0.139438	2	0.7	0	0		0.0003	2	8.33	11.33
1,1,2-dichloroethane	0.00135	0.00135	0.0014	0	0	1		0	0		0.0003	1	4.17	4.50
Tetrachlorethylene	0.0034	0.0034	0.0034	0	0	1	0.04	0	0		0.0002	1	4.17	17.00
Chlorobenzene	0.00014	0.31	0.0963	0.0769	79.86968	5	0.3	0	0		0.0001	5	20.83	3,100.00
Ethylbenzene	0.128	0.128	0.128	0	0	1	0.3	0	0		0.0003	1	4.17	426.67
P-xylene	0.00164	0.00164	0.0016	0	0	1	0.5	0	0		0.0003	1	4.17	5.47
O-xylene	0.00234	0.00234	0.0023	0	0	1	0.5	0	0		0.0003	1	4.17	7.80
1,3-dichlorobenzene	0.0019	0.00282	0.0024	4E-07	0.017932	2		0	0		0.0001	2	8.33	28.20
1,4-dichlorobenzene	0.00216	0.00543	0.0038	5E-06	0.140881	2	0.3	0	0		0.0001	2	8.33	54.30
1,2-dichlorobenzene	0.0004	0.00801	0.004	3E-05	0.722302	3	1	0	0		0.0001	3	12.50	80.10
1,2,4-trichlorobenzene	0.0016	0.00958	0.0056	3E-05	0.569592	2	0.02	0	0		0.0002	2	8.33	47.90

Note 1. Background values and limits of detection are used as comparison standards for inorganic and organic components respectively
2. The background values of groundwater in the study area are drawn from the natural background values of groundwater in the Second Songhua River basin identified in the 1988 study on Chemical Background Values of Groundwater in Jilin Province. Among them, the NO$_3$–N and salinity background values use the initial groundwater pollution values of the Jilin city identified in the Study on Urban and Suburban Groundwater Quality, Cause and Protection in Jilin City conducted in 1983
3. Class III groundwater quality standards draw reference from the *Environmental Quality Standards for Groundwater* (2009): Draft for Approval

4.2 Groundwater Quality Evaluation

Based on the results of groundwater sampling analysis, the study selects a total of 22 inorganic indicators (pH, TH, TDS, SO_4^{2-}, Cl^-, NO_3–N, NO_2–N, NH_4–N, COD, I, F, Hg, As, Cr^{6+}, Cd, Pb, Zn, Al, TFe, Mn, phenolics and cyanide) and 24 organic indicators (vinyl chloride, 1,1-dichloroethylene, 1,2-dichloroethylene, chloroform, carbon tetrachloride, 1,2-dichloroethane, benzene, trichlorethylene, 1,2-dichloropropane, toluene, 1,1,2-dichloroethane, tetrachlorethylene, chlorobenzene, ethylbenzene, p-xylene, o-xylene, 1,3-dichlorobenzene, 1,4-dichlorobenzene, 1,2-dichlorobenzene, 1,2,4-trichlorobenzene, total BHC, hexachlorobenzene, total DDT, and BaP). First, individual factors are evaluated and then, comprehensive groundwater quality evaluation is conducted using the Nemerow's Index. The specific steps are described below.

(1) Individual factor evaluation

Groundwater components are divided to five categories according to Table 4.1 "Conventional Indicators and their Limits for Groundwater Quality Standards" and Table 4.2 "Unconventional Indicators and their Limits for Groundwater Quality Standards" of the GB T 14848-2007 *Groundwater Quality Standards*. The results of individual factor evaluation are as shown in Figs. 4.2 and 4.3. According to the results about inorganic components, the groundwater quality reaches Class V in most parts and Class IV in certain parts, and the chemical components influencing the groundwater quality in the study area include NO_3–N, TFe, Mn, Hg, NH_4–N, TDS, SO_4^{2-} and Al^{2+}. The results about organic components indicate that the groundwater samples reach Class I, III, IV and V, accounting for 25%, 54.17%, 16.67% and 4.17% respectively. BaP and carbon tetrachloride are mainly responsible for Class IV quality and benzene for Class IV quality.

(2) The scores of individual factors Fi are obtained according to requirements as shown in Table 4.3.
(3) The comprehensive score F is calculated using Formulas (4.1) and (4.2)

$$F = \sqrt{(\bar{F}^2 + F_{\max}^2)/2} \qquad (4.1)$$

$$\bar{F} = \frac{1}{n}\sum Fi \qquad (4.2)$$

Wherein $\bar{F}$ represents the average of the scores of individual components Fi; F_{Max} represents the maximum of the scores of individual components Fi, and n stands for the number of items.

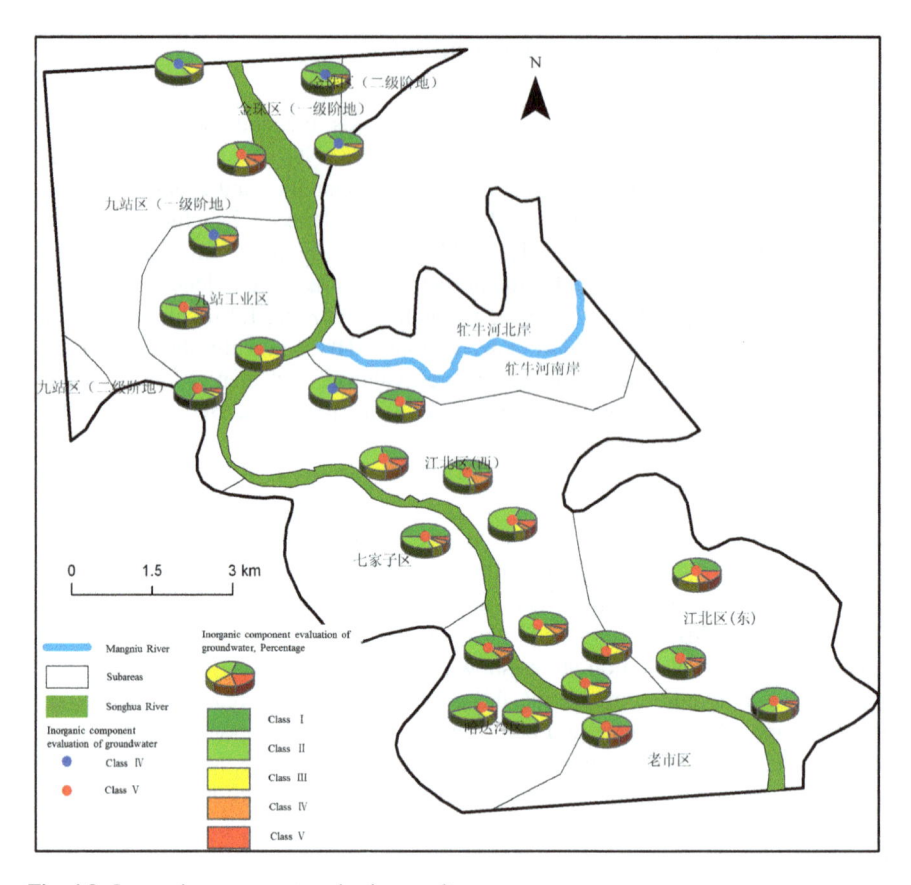

Fig. 4.2 Inorganic component evaluation results

(4) Groundwater quality is graded according to the obtained scores F_i, as shown in Table 4.4.

The results of comprehensive groundwater quality in the study area are as shown in Fig. 4.4. The groundwater has poor quality in vast majority of the area with extremely poor quality observed in the diphenyl plant of Jilin Chemical Company and the surrounding area (JL1011).

4.3 Groundwater Pollution Assessment

Based on the results of groundwater sampling analysis, the pollution of individual factors is first evaluated, including 10 inorganic components (Cl⁻, SO_4^{2-}, NO_3–N, TH, TDS, COD, mercury, arsenic, chromium, lead) and 12 organic components (BaP, toluene, trichlorethylene, benzene, chlorobenzene, vinyl chloride,

4.3 Groundwater Pollution Assessment

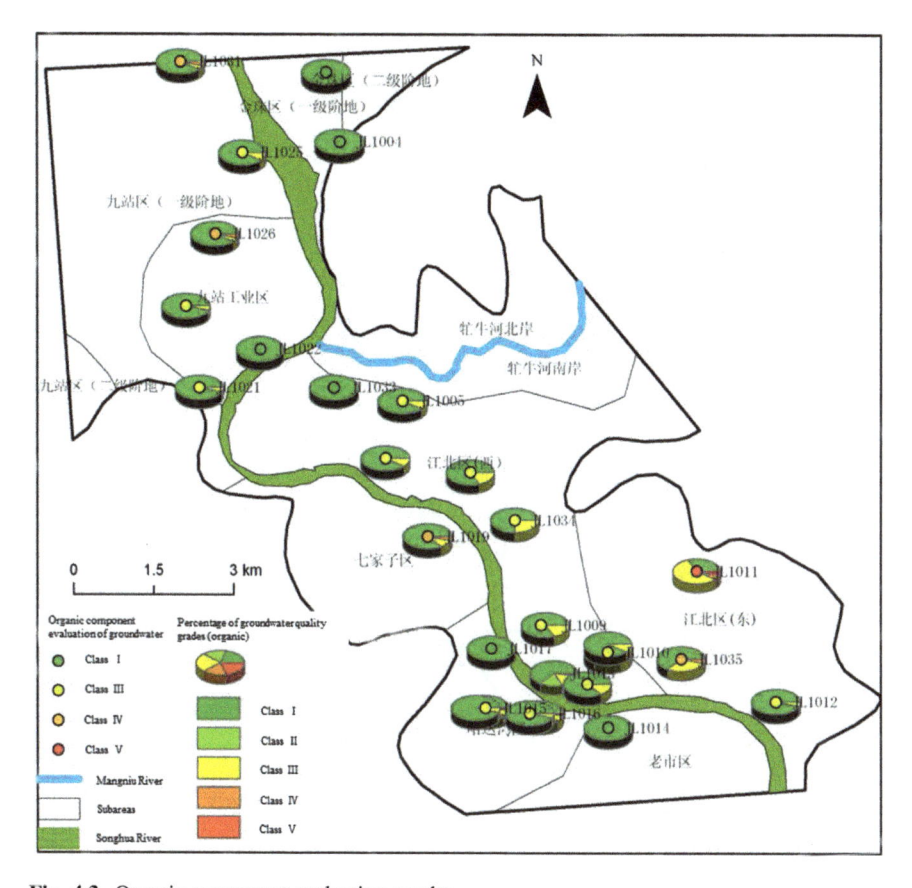

Fig. 4.3 Organic component evaluation results

Table 4.3 Individual factor score by group

Group	I	II	III	IV	V
Fi	0	1	3	6	9

Table 4.4 Groundwater quality scores in different group

Group	Excellent	Good	Moderate	Poor	Very poor
Fi	F < 0.8	$0.8 \leq F < 2.5$	$2.5 \leq F < 4.25$	$4.25 \leq F < 7.20$	$F \leq 7.20$

1,2-dichloroethylene, carbon tetrachloride, 1,2,4-trichlorobenzene, ethyl benzene, total DDT). The overall groundwater pollution is assessed using the superposition grading index method. The specific steps are described below.

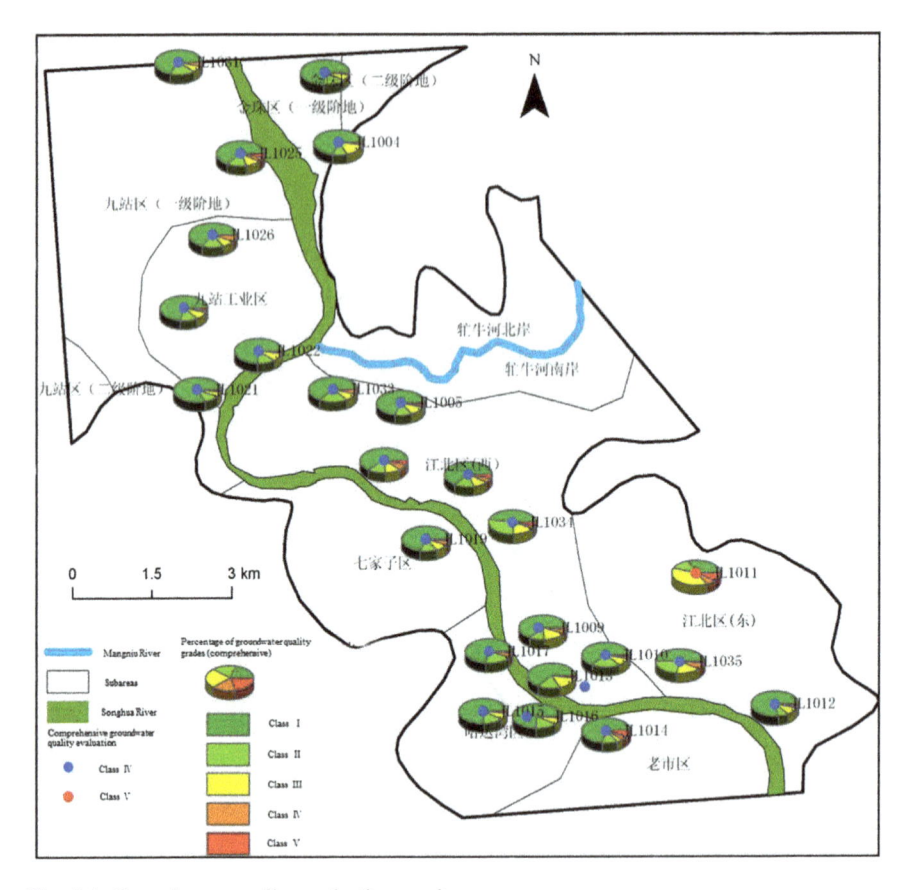

Fig. 4.4 Groundwater quality evaluation results

(1) The pollution index of individual factors *I* is calculated using Formula (4.3) and compared with Table 4.5 to visually observe the degree of component-specific pollution.

$$I = \frac{C_i}{C_{oi}} \qquad (4.3)$$

Wherein *I* represents the pollution index of individual factors, dimensionless; C_i is the measured concentration of component *i*, mg/l and C_{oi}, control value of component *i*. The criteria of assessment are groundwater quality control values of this study area (Table 4.6) for inorganic pollutants and limits of detection in standard detection (Table 4.2) for trace organic pollutants.

4.3 Groundwater Pollution Assessment

Table 4.5 Pollution index of individual factors compared with component-specific groundwater pollution

I ≤ 1	No pollution	Below the control values or background values
1 < I ≤ 5	Light pollution	1–5 times more than the control values or background values
5 < I ≤ 10	Moderate pollution	5–10 times more than the control values or background values
10 < I ≤ 50	Heavy pollution	10–50 times more than the control values or background values
I > 50	Severe pollution	50 times more than the control values or background values

(2) Superposition grading index

Scores are given according to the pollution index of individual factors I, as shown in Table 4.7.

The pollution index PI of each groundwater sample is calculated using Formula (4.4).

$$PI = \sum_{i-1}^{n} Fi \qquad (4.4)$$

Then, groundwater pollution is graded according to Table 4.8.

According to the above steps, the results of groundwater pollution assessment in the study area are drawn, as shown in Fig. 4.5.

Groundwater in the study area is neither unpolluted nor slightly polluted, but moderate pollution, heavy pollution and severe pollution are identified.

Moderate pollution is founded in 4 wells, accounting for 16.67% of the total. These wells are distributed in the vast farming area in the west and north, and partially south of Jiuzhan Economic Development Zone, farming area of Jinzhu County and Wulajie Town, Songyuan Hada water source area in the north of Mangniu River, and farming area of Qipan Village and Tiedong Village in the east. Due to excessive application of urban faeces, manure, and nitrogen fertilizers in agriculture beyond the absorption capacity of plants, decomposition products and nitrogen are retained in the soil and penetrate into the ground through rain, contaminating the groundwater. Thus, the major source of groundwater pollution is agriculture, with such pollutants as NO_3–N, SO_4^{2-} and Cl^-.

Another 14 wells, accounting for 58.33%, are graded heavily polluted. They are in the Jiuzhan Economic Development Zone and its villages, northern petrochemical industrial zone along Xiweizi-JCGC Chemical Refinery and surrounding old residential areas, northern part of the old area of Jilin City, Hadawan Industrial Zone and the northern Qijiazi vegetable area. As the industrial area exists for nearly

Table 4.6 Determination of chemical background values and control values of groundwater in the study area

Item/Value	K⁺	Na⁺	Ca²⁺	Mg²⁺	Cl⁻	SO₄²⁻	NO₃⁻-N	HCO₃⁻	Fe	Mn	pH
X	1.558	21.287	47.310	12.941	12.399	2.889	1.53	241.814	3.674	0.980	7.06
S	0.720	0.504	21.497	0.506	7.326	1.110		0.024	1.460		0.370
X + 2S	2.998	22.295	90.300	13.922	27.051	5.109	1.53	242.562	6.594		7.80
Item/Value	COD	F	I	TH	TDS	Pb	Zn	Hg	Cd	Cr⁶⁺	As
X	1.267	0.354	3.726	159.84		0.338	12.653	0.020	0.046	0.666	0.788
S	0.608	0.178	6.660	9.22		0.794	11.360	0.740	0.984	1.280	0.962
X + 2S	2.483	0.710	0.005	179.28	104.712	0.0019	0.0015	0.0015	0.0020	0.0032	0.003

Note 1. The mean background values and standard deviations of I, Pb, Zn, Hg, Cd, Cr⁶⁺, and As are expressed by μg/l, and the rest, mg/l
2. X is the background value, S, standard deviation, X + 2S control value, expressed by mg/l

4.3 Groundwater Pollution Assessment

Table 4.7 Pollution index of individual factors

I	I ≤ 1	1 < I ≤ 5	5 < I ≤ 10	10 < I ≤ 50	I > 50
Fi	1	10^2	10^4	10^6	10^8

Table 4.8 Groundwater pollution grades

PI range	Grade	Explanation
1 < PI ≤ 100	No pollution (I)	Below the control values of all components
$100 < PI \leq 10^4$	Light pollution (II)	1–5 times more than the control values of one or more components
$10^4 < PI \leq 10^6$	Moderate pollution (III)	5–10 times more than the control values of one or more components
$10^6 < PI \leq 10^8$	Heavy pollution (IV)	10–50 times more than the control values of one or more components
$PI > 10^8$	Severe pollution (V)	Over 50 times more than the control values of one or more components

50 years, industrial sewage containing inorganic and organic compounds leak underground due to disrepair of drainage system. Volatile organic components and Hg diffuse into the atmosphere and by dry and wet deposition, land and seep into the ground with rain. In addition, domestic sewage, faecal water, and domestic filtration water in old residential communities (urban cottages) and villages filtrate underground through seepage wells, permeable toilets and garbage stacks. All these processes severely pollute the groundwater. In general, domestic and industrial sources have a dominant role in severe groundwater pollution, involving such pollutants as NO_3–N, Cl^-, SO_4^{2-}, TH, BaP, benzene and its homologues, and halogenated aliphatic hydrocarbons—alkanes, olefins, and total DDT.

Groundwater pollution is very heavy in 6 wells, accounting for 25% of the monitored ones. These wells are located in the northern Petrochemical Industrial Zone with JCGC diphenyl plant as the center, which extends to JCGC pesticide plant in the north JCGC organic synthesis plant in the east, and central Jilin thermal power plant in the south, and also located in the area east and southwest of the open channel of Jiuzhan Industrial Zone. The pollution is mainly caused by the years-long leakage and diffusion of industrial sewage containing organic compounds in the industrial zone and industrial sewage containing SO_4^{2-} and TDS in the open channel, which collectively seen as industrial sources. The pollution factors mainly include NO_3–N, SO_4^{2-}, TH, TDS, BaP, benzene and its homologues, and halogenated aliphatic hydrocarbons—alkanes, olefins.

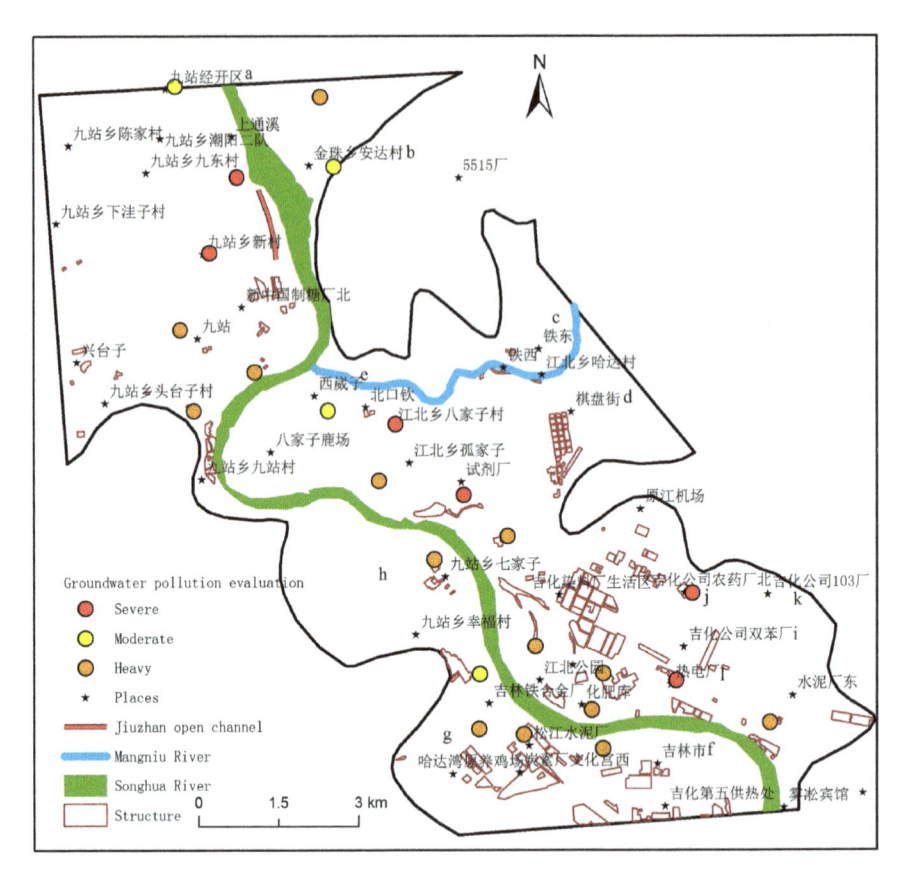

Fig. 4.5 Groundwater pollution assessment results. Note: a. Jiuzhan Economic Development Zone; b. Jinzhu County; c. Tiedong Village; d. Qipan Village; e. Xiweizi; f. the old area of Jilin City; g. Hadawan Industrial Zone; h. Qijiazi area; i. JCGC diphenyl plant; j. JCGC pesticide plant; k. JCGC organic synthesis plant; l. Jilin thermal power plant

4.4 Groundwater Pollution Source Apportionment

The investigation into groundwater pollution of Songhua River (Jilin Section) conducted in 2005 covers the survey of pollution sources and interpretation of land use types using remote sensing. Therefore, this study analyzes the groundwater pollution sources, taking into account land use, source distribution and spatial distribution of excessive components.

4.4 Groundwater Pollution Source Apportionment

4.4.1 Groundwater Pollution Sources and Land Use Types

Figure 4.6 shows the spatial distribution of groundwater pollution sources according to land use types, wherein the sewage outfalls along the Songhua River are mainly for industrial purpose. The details are presented as follows:

(1) Industrial sources

A. Industrial sewage outfalls

The main sources of groundwater pollution in the city of Jilin are distributed on both sides of the Songhua River. A variety of contaminants enter surface water through sewage outlets on the two sides. Given that surface water and groundwater in the study area have close hydraulic connection and frequent exchange, these contaminants flow from the river into the groundwater with replenishment in the

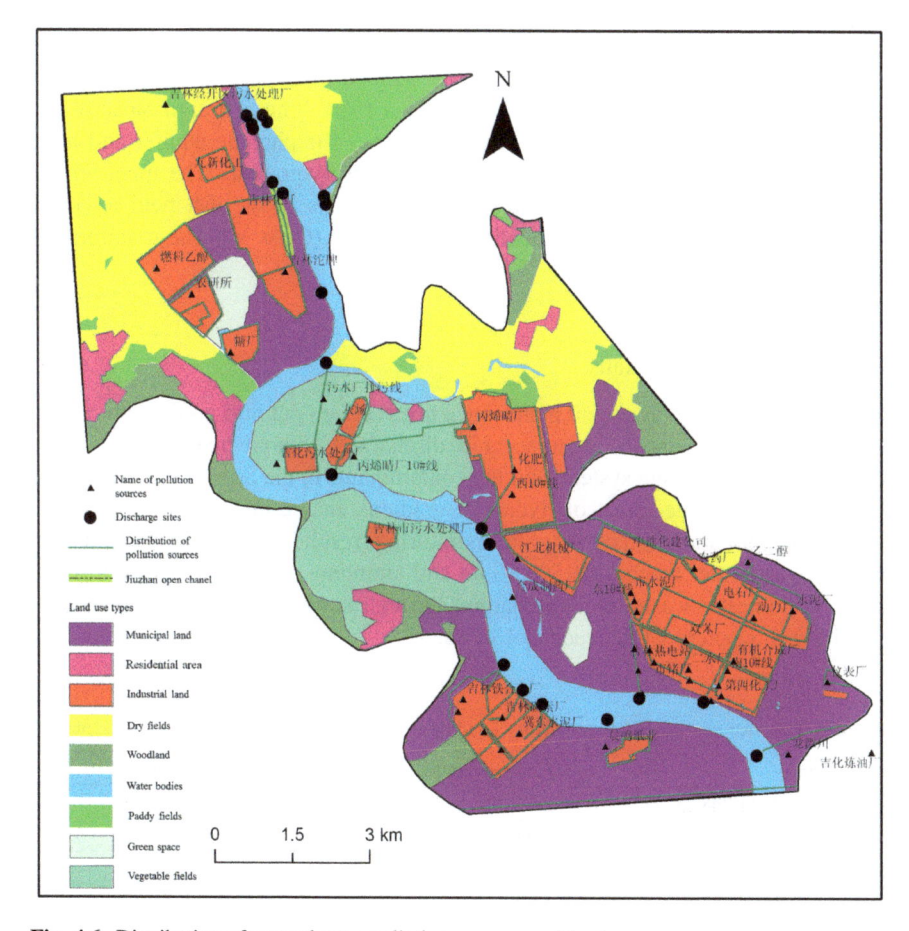

Fig. 4.6 Distribution of groundwater pollution sources and land use

wet season and lead to groundwater pollution under the effect of hydraulic gradient. In 2005, totally 13,711.8 × 10⁴ t/a of sewage was discharged by industrial enterprises to the Songhua River through the outfalls for industrial wastewater and mixed wastewater. Among them, 8889 × 10⁴ t/a met the standards after treatment; 4,744.8 × 10⁴ t/a was directly discharge to the river, and 78 × 10⁴ t/a exceeded the standards. The COD emissions from major industrial enterprises were up to 6685.1t/a and industrial solid wastes to 550.86 × 10⁴ t/a.

B. Industrial solid waste

Jilin is an industrial city underpinned by chemical, electric power, metallurgy and mining sectors and has a coal-dominated energy structure. Industrial solid waste produced in the city is primarily composed of fly ash, cinder, slag, gangue, and tailings. Among them, fly ash mainly comes from Jilin and Xinli thermal power plants fired by pulverized coal and fuel ethanol enterprises; cinder from coal-fired boilers excluding electric boilers; slag from ferroalloy plants and other smelting enterprises. In 2005, the industrial solid waste amounted to 550.86 × 10⁴ t/a, including 110.88 × 10⁴ t/a of slag and 156.92 × 10⁴ t/a of fly ash.

The ash field of JCGC fertilizer plant was located in Bajiazi Village, Jiangbei County, Longtan District, and put into operation in 1994. It has a design capacity of 265 × 10⁴ m³ and design life of 10 years. It reaches the phreatic aquifer and has impermeable layers in the inner side of dam and reservoir base. The monitoring results show that groundwater in the surrounding areas of the ash field is contaminated by fluoride and poor in quality, if not worse (Class V). The monitoring results are detailed in Table 4.9.

(2) Domestic sources

A. Municipal sewage outfalls

Jilin municipal sewage treatment plants have not yet been put into production. As a result, 78% of the domestic wastewater is discharged without treatment into the river through the drainage system, and the contaminants in the Songhua River enter groundwater under the effect of hydraulic gradient, increasing the possibility of groundwater pollution. According to monitoring results, the COD content of domestic wastewater reaches 21,506.86 t/a and ammonia nitrogen 1,701.82 t/a. In 2005, the domestic wastewater discharged through sewage outfalls amounted to 61 × 10⁴ t/a, wherein COD reached 136 t/a.

Table 4.9 Downstream groundwater quality conditions of the ash field of JCGC fertilizer plant

Item Location	Fluoride content (mg/l)	Quality level
South of ash field 248 m	2.15	Extremely poor
West of ash field 200 m	2.92	Extremely poor
Southwest of ash field 50 m	2.80	Extremely poor
West of ash field 713 m	1.80	Extremely poor

4.4 Groundwater Pollution Source Apportionment

Table 4.10 Monitoring results of main contents of Longtanqu landfill leachate in 2004

Item	COD	Volatile phenol	Ammonia nitrogen	Nitrate
Content(mg/l)	5,965.7	1.9	2,086.25	25.6

B. Municipal solid waste and excrement

In Jilin, municipal solid waste is mainly landfilled. Currently, there are three sanitary landfills, i.e. the Municipal Landfill, High-Tech New District Landfill and Longtan District Landfill. Reaching the end of service, the Longtan District Landfill has accepted nearly 50×10^4 tons of garbage from the Longtan District and JCGC. Without garbage dams, the landfill leachate discharges naturally and pollutes the soil and groundwater of downstream valley of stacks. The downstream groundwater quality has exceeded the Class IV standards of Groundwater Quality Standards (GB/T14848-93) due to such pollutants as COD, ammonia nitrogen, nitrate and volatile phenol. Table 4.10 shows the pollutant contents of landfill leachate according to the 2004 dataset of Jilin Environmental Monitoring Station.

(3) Agricultural sources

In the study area, there are Longtan Township, Jiangbei Township and Jinzhu Township, with land mainly used as paddy fields, dry fields and vegetable fields. Chemical fertilizers and pesticides are applied in 78.5% of the land. In specific, diammonium is applied in holes in May, urea in June; special fertilizer for rice in May; compound base fertilizer in May–July; and herbicide and insecticides sprayed in June–July for paddy fields, and in May for dry land.

(4) Polluted surface water bodies

Located in the east of Jiuzhan industrial zone, the open channel is originally natural watercourses with sand and gravel in the riverbed and run for many years without seepage control. Prior to 1992, it was used to discharge industrial wastewater from sugar refineries and chemical plants. Fiber and lignin, whose contents are high in the wastewater, gradually silted up sewers, reducing the leakage of sewage to a certain extent. With the shutdown of sugar refineries and operation of new chemical projects after 1992, the open channel served to discharge industrial wastewater from Tianhe alcohol plant, domestic seweage from sugar mills, and industrial and domestic wastewater from chemical fiber plants. Currently, the 1,600 m long and 3–5 m wide open channel has a flow of $2,500 \times 10^4$ m^3/a. According to 2005 monitoring data (Table 4.11), the groundwater quality index of this plot ranges from 4.26 to 7.16, which represents poor water quality and heavy water pollution. Groundwater here has basically been inappropriate for drinking.

Table 4.11 Water quality test results of Jiuzhan open channel in 2005 (unit: mg/l)

Item No.	Total iron	Total manganese	Ammonia nitrogen	Chloride	Sulfate	Nitrate nitrogen	Nitrite nitrogen	Fluoride	Total phosphorus
1	0.46	3.29	4.88	<DL	401	0.90	0.012	1.16	0.01
2	1.00	2.17	8.78	25.5	248	0.75	0.033	0.73	0.01
Pollutant No.	TH	Salinity	COD	Volatile phenol	Total arsenic	Cadmium	Hexavalent chromium	PH value	Conductivity
1	679	298.6	39.6	0.003	<DL	0.0036	<DL	6.9	1,000
2	889	759.1	17.0	0.004	<DL	0.0009	<DL	6.8	1,350

4.4.2 Groundwater Pollution Source Apportionment

Groundwater pollution in the study area can be categorized into industrial source pollution, agricultural source pollution and domestic source pollution. It can be caused by point sources (industrial and domestic outlets, industrial solid waste, domestic waste and manure), linear sources (Jiuzhan open channel), and non-point sources (fertilizers and pesticides in paddy fields, dry fields and vegetable fields). The chemical pollutants include COD, NO_3–N, NH_4–N, TFe, Mn, fluoride, TDS, SO_4^{2-}, Al, heavy metals (Hg, Zn and Cr), organic matters (volatile phenols, BaP, carbon tetrachloride, benzene, chlorobenzene), and the like.

Pollutants and pollutant types, non-attainment areas (also illustrated in Fig. 4.7) and pollution sources (Table 4.12) are summarized below, based on the test results of groundwater samples collected during the normal period in 2005 and chemical components of solid waste and open channel identified through pollution source surveys.

(1) SO_4^{2-}, TH and TDS

In the study area, the groundwater concentrations of SO_4^{2-}, TD and TDS are relatively high in riverside rural residential areas, which is related to the widespread use of non-seepage toilets, heaps of human and animal manure around houses, and sewage wells (pits). At the same time, the entry of Ca^{2+} and Mg^{2+} into groundwater through exchange with Na^+ contained in the sewage via soil increases the TH level. Further, increased macro components and TH allow for higher TDS content. In addition, through the open channel on the east side of Jiuzhan water source, industrial effluent with high SO_4^{2-} and TDS content is discharged into the ground, resulting in belt groundwater pollution on both sides.

(2) NO_3–N and NH_4–N

The nitrate sources in groundwater pollution in the study area include (a) agricultural non-point sources: The contamination of groundwater below the vegetable fields is most serious because of more fertilizer and pesticide application and recharge to shorten the growth period and increase the yield; (b) industrial wastewater, waste gas and solid waste: The three wastes from industrial enterprises penetrate underground through the vadose zone driven by precipitation, surface water, irrigation, or directly pollutes groundwater as a result of leakage through seepage wells, pits, and pipelines; (c) domestic garbage: A huge amount of sewage and garbage are produced in populous areas, such as Old Urban District, Jiangbei, Hadawawa and Qijiazi. The inappropriate storage and imperfect drainage system leads to nitrogen infiltration and groundwater pollution.

The groundwater contaminant of ammonia nitrogen in Jiuzhan District and Jinzhu District is mainly attributed to the use of agricultural pesticides and fertilizers, while in Jiangbei County and Jiangbei Petrochemicals Zone on the second-order terrace, is primarily affected by industrial sources, including chemical manufacturing, refrigerants, synthetic fibers and fuels. In Shijiazi, such contamination is mainly due to domestic garbage.

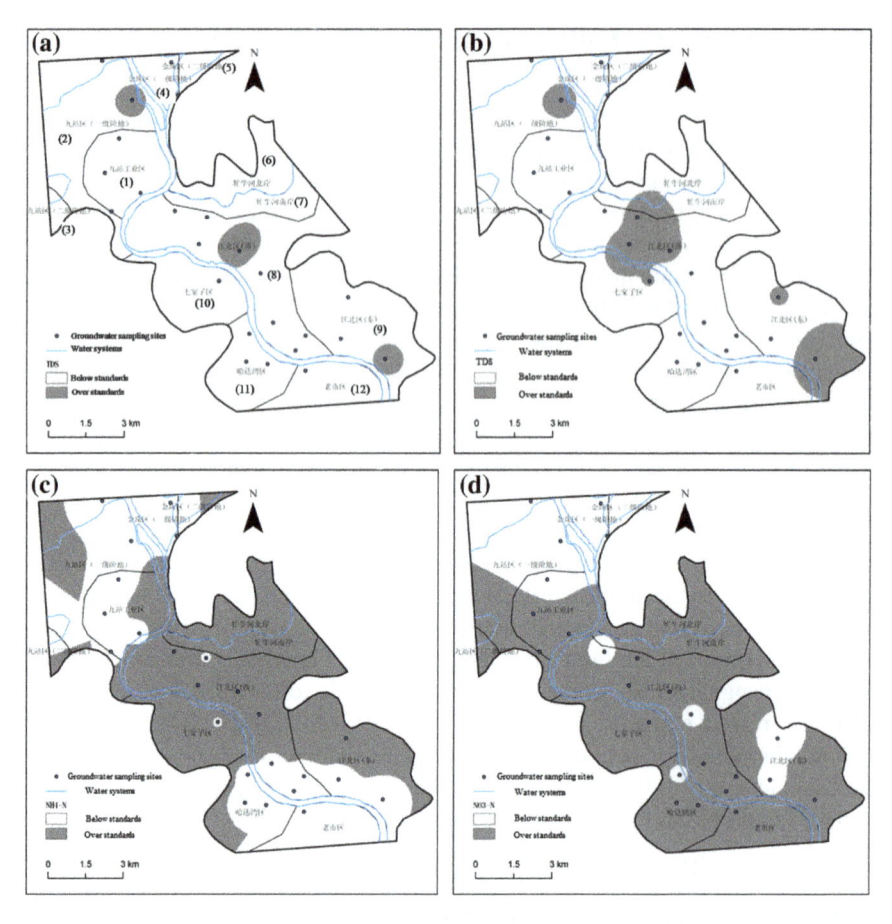

Fig. 4.7 Spatial distribution of standard-exceeding groundwater components. Note: (1) Jiuzhan Industrial Zone; (2) Jiuzhan District (first-order terrace); (3) Jiuzhan(second-order terrace); (4) Jinzhu District(first-order terrace); (5) Jinzhu district (second-order terrace); (6) North bank of Mangniu River; (7) South bank of Mangniu River; (8) (western) Jiangbei District; (9) (eastern) Jiangbei District; (10) Qijiazi District; (11) Hadawan District; (12) Old urban District

(3) TFe and Mn

TFe and Mn exceeded the groundwater standards in four parts of the study area: vicinity of Shangtong River near the Songhua River floodplain, Anda Village of Jinzhu County, Thermal Power Plant—Pesticide Factory—Bajiazi Village in Jiangbei County, and floodplain and first-order terrace in Hadawan. In general, iron (manganese) is rich in valley floodplains and first and second-order terraces of the Songhua River due to special hydro-geochemical environment. These components cause serious damage to groundwater quality, which is related to such factors as good closed reduction environment, rich organic matter in aqueous media and stagnant groundwater flow.

4.4 Groundwater Pollution Source Apportionment

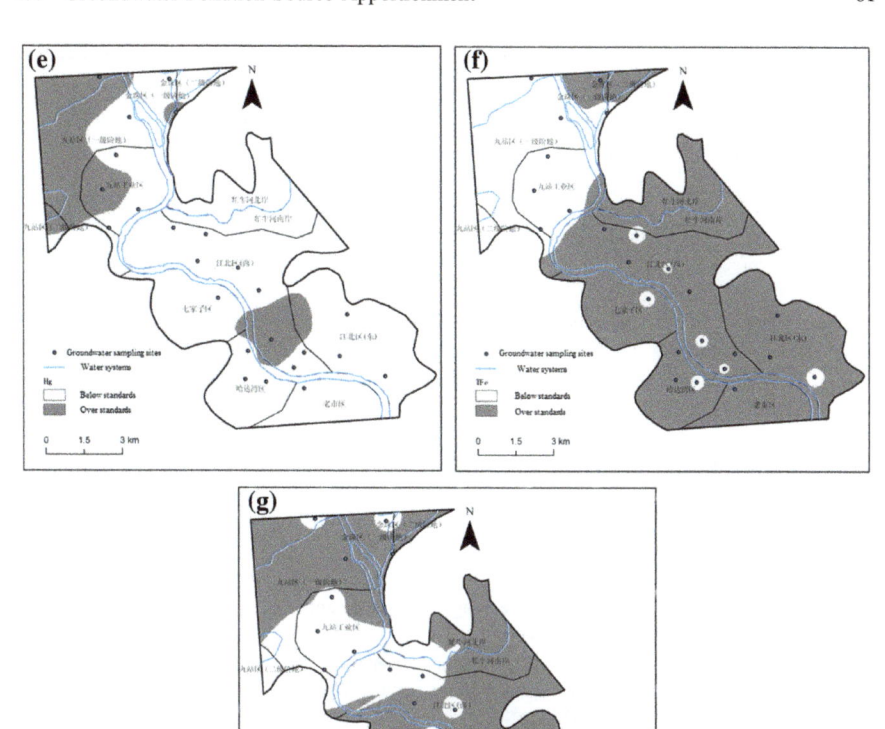

Fig. 4.7 (continued)

(4) Hg

Excessive groundwater mercury is found in west and southwest parts of Jiangbei Petrochemical Industrial Zone, Qijiazi vegetable fields in the north of Hadawan Industrial Zone, some wells in Jiuzhan Industrial Zone and the northeast areas, as well as Anda Village of Jinzhu County. It is probably because mercury-containing volatile particulate matter in the exhaust gas of heavy industries falls to the ground through dry and wet deposition and accumulates in the soil over years. As a result, mercury enters and pollutes groundwater through recharge by surface water. The heavy industries include petrochemical industry in Jiangbei, paper, cement, carbon, iron alloy industries in Hadawan, and the industry in Jiuzhan.

Table 4.12 Groundwater pollutant types in the study area, non-attainment areas and pollution sources

Type	Pollutants	Distribution	Sources
Integrated indicators	TDS	Figure 4.7a	Industrial, domestic sources
	COD	Old urban areas along the left bank	Industrial, drainage rivers
	TH	Figure 4.7b	Industrial, agricultural, domestic sources and drainage rivers
Inorganic matters	SO_4^{2-}	Hadawan Area, Jiuzhan open channel	Domestic sources, drainage rivers
	Fluoride	Ash field of Bajiazi Village of Jiangbei County, and open channel of Jiuzhan Industrial Zone	Industrial, drainage rivers
	NH_4-N	Figure 4.7c	Industrial, agricultural, drainage rivers
	NO_3-N	Figure 4.7d	Industrial, agricultural, domestic sources
Heavy metals	Hg	Figure 4.7e	Industrial
	Cr	Outfalls along the Songhua River (Jiangbei Machinery Plant)	Industrial
	Zn	Outfalls along the Songhua River (Jiangbei Machinery Factory, Jilin Chemical Fiber Co., Ltd.)	Industrial
Others	TFe	Figure 4.7f	Native
	Mn	Figure 4.7g	Native
Organic matters	BaP	Jilin thermal power plant in (eastern) Jiangbei District, Jiuzhan Industrial Zone, and Wastewater treatment plant in Jilin Economic Development Zone in Jiuzhan first-order terrace	Industrial, agricultural
	Carbon tetrachloride	Vegetable fields in Qijiazi	Agricultural
	Benzene	JCGC diphenyl plant in (eastern) Jiangbei District	Industrial
	Volatile phenol	Open channel of Jiuzhan Industrial Zone	Industrial

(5) Organic compounds

Among a variety of petrochemical contaminants detected in the study area, the organic compounds that exceed the standards mostly concentrated in the 50-year-old petrochemical industrial zone. Pollutants leak underground as dense drainage stems and lines are not well maintained and repaired over the years. Irrational industrial wastewater discharge is first to blame in the such chemical pollution of groundwater.

Chapter 5
Groundwater Pollution Risk Assessment

Abstract Groundwater pollution prevention and control should be given priority to avoid costly remediation since groundwater pollution is invisible, complex, and has long-term impact. Groundwater pollution risk assessment, which refers to the process of determining potential impacts of any pollutant, is an effective tool for designing efficient groundwater management and protection strategies. The chapter adopts the index overlay method to assess the risk of groundwater pollution covering a variety of typical pollutants in the study area. Groundwater pollution risk is believed to be a superposed result of intrinsic groundwater vulnerability and groundwater pollution load, and groundwater function, so the risk assessment is focused on pollution load assessment and groundwater function assessment.

Keywords Intrinsic groundwater vulnerability · Pollution load
Groundwater function · ArcGIS-based matrix method · Level difference method
Groundwater pollution risk assessment

5.1 Intrinsic Groundwater Vulnerability Assessment

Intrinsic groundwater vulnerability reflects the self-purification capacity of groundwater systems to absorb pollutants and to a certain degree, the speed and quality that contaminants reach aquifers. The DRASTIC model used to assess intrinsic groundwater vulnerability sets seven hydrogeological parameters, including depth to water table (D), net recharge (R), aquifer media (A), soil media (S), topography (slope, T), impact of vadose zone media (I) and hydraulic conductivity (C). Each indicator is divided into several levels, each of which is given a score ranging from 1 to 10. The influence of these indicators on vulnerability is weighted (5, 4, 3, 2, 1, 3 and 5) and added using Formula (5.1) to obtain the groundwater vulnerability index referred to as the DRASTIC Index (DI) (Aller 1987). According to this index, vulnerability is graded, such as low, medium and high. A great DI value indicates high vulnerability, and vice versa.

$$DI = D_W D_R + R_W R_R + A_W A_R + S_W S_R + T_W T_R + I_W I_R + C_W C_R \qquad (5.1)$$

Wherein the subscript R represents the value of indicators and the subscript W the weight of indicators.

Using ArcGIS, the rating layer of the seven indicators is drawn based on topographic maps (1: 50,000), geologic maps, hydrogeological maps, drilling histograms, meteorological and hydrological data, and field sampling and laboratory test results provided by the Geological Environment Monitoring Station of Jilin Province. The data sources for each parameter are specified in Table 5.1.

According to the level, score and weight of seven indicators developed by Aller (1987), the groundwater vulnerability of the study area is evaluated. The rating layers by indicators are as shown in Fig. 5.1.

The assessment results show that high groundwater vulnerability is seen in the floodplain of the Mangniu River and first-order terrace of Anda Village in the northern Jinzhu County and sporadically in the first-order terrace on the left bank of the Second Songhua River. Groundwater vulnerability is relatively high in the first-order terrace of the Mangniu River and floodplain of the Second Songhua River, as well as sporadically in the vicinity of villages of Jiuzhan County and Gujiazi and Bajiazi of Jiangbei County. Groundwater is moderately vulnerable in the first-order and second-order terraces of the two sides of the Second Songhua River. The low-vulnerability zone covers the rear edge of the first-order terrace and second-order terrace, as well as sporadic areas along the right bank of the Second Songhua River in the southern part of the study area, such as Jiangbei Park, Fertilizer Warehouse and Cement Plant East. Both moderate and low vulnerability are found in the floodplain along Thermal Power Plant and Wusong Hotel on the right bank of the Second Songhua River, second-order terrace west of Toutaizi Village in Jiuzhan County, as well as first-order terrace east of Toutaizi Village in Jiuzhan County.

Table 5.1 DRASTIC model data sources *Source* The table is reprinted from (Huan et al. 2018), with permission from Elsevier

Parameter	Data source
D	34 monitoring sites of shallow groundwater table in the normal season in 2005
R	*Reports on Dynamic Groundwater Monitoring in Jilin* (1991–1995, 1997) and *Jilin Hydrogeological Investigation Report* (2008)
A	84 drilling histograms (aqueous layer lithology)
S	Particle analysis of 28 soil samples collected in 2011
T	84 drilling histograms (surface elevation)
I	84 drilling histograms (lithology)
C	*Reports on Dynamic Groundwater Monitoring in Jilin* (1991–1995, 1997) and *Jilin Hydrogeological Investigation Report* (2008)

5.2 Pollution Load Assessment

Pollution load assessment measures the possibility and load of groundwater pollution caused by human activities and various sources. Based on the collected data of pollution sources, quantitative evaluation is conducted to grade the comprehensive pollution load of typical pollutants from different sources. More specifically, the risk of pollution sources is rated according to the category, quantity, storage, and self-purification of pollutants, as well as the migration to groundwater.

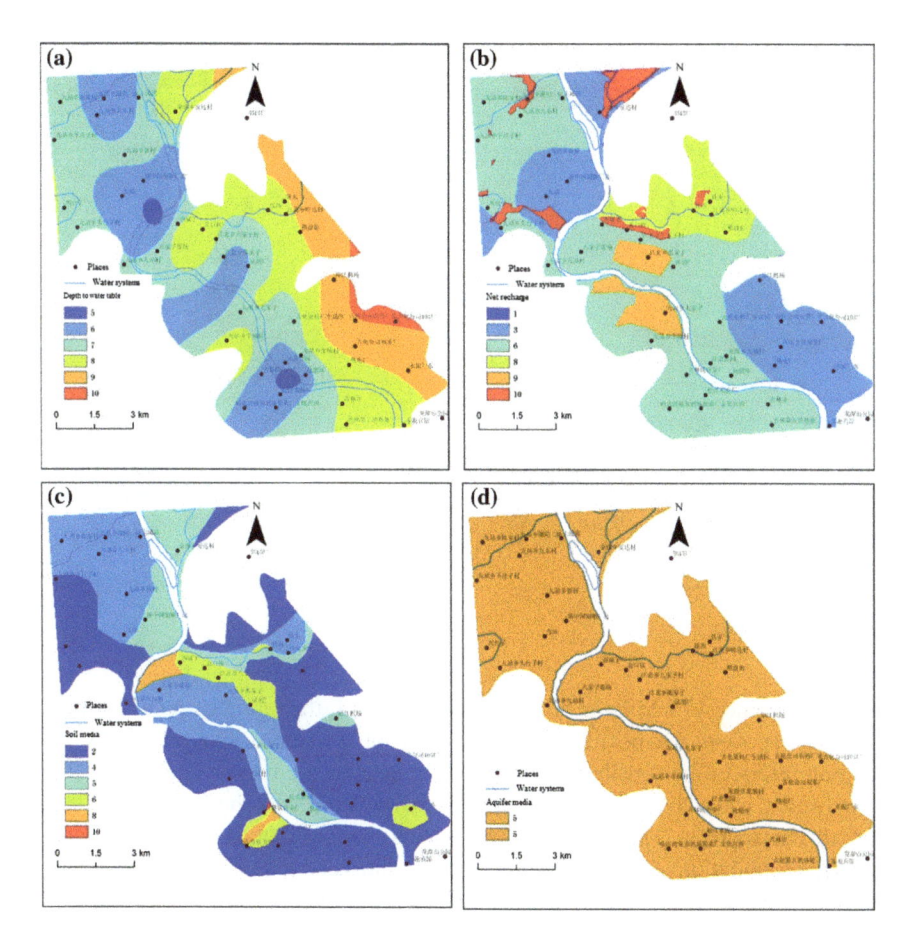

Fig. 5.1 DRASTIC model indicator scores and vulnerability assessment results. Note: **A** Depth to water table; **B** Net recharge; The figure is reprinted from (Huan et al. 2016), with permission from Elsevier. **C** Soil media; The figure is reprinted from (Huan et al. 2016), with permission from Elsevier. **D** Aquifer media; **E** Topography (slope); **F** Impact of vadose zone media; **G** Hydraulic conductivity; **H** DRASTIC evaluation results. In Figure (**H**), (a) Anda Village of Jinzhu Country; (b) Jiuzhan County; (c) Gujiazi; (d) Bajiazi; (e) Jiangbei Park; (f) Fertilizer Warehouse; (g) Cement Plant East; (h)Thermal Power Plant; (i) Wusong Hotel; (j) Toutaizi Village. The Figure (**H**) is reprinted from (Huan et al. 2018), with permission from Elsevier

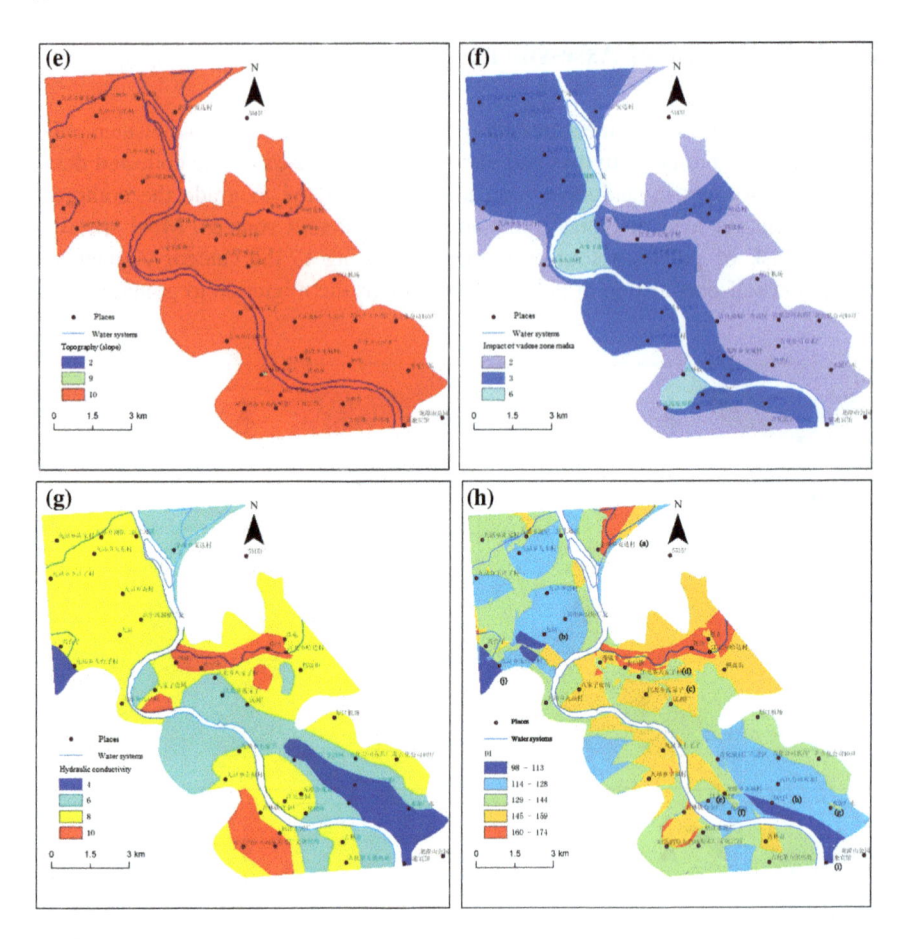

Fig. 5.1 (continued)

With reference to USEPA's priority-setting method, the rating of pollution risk takes into account the possibility and severity of pollution. The possibility of pollution refers to the possibility that pollution sources are released and reach the groundwater, while the severity of pollution mainly considers the nature of pollutants in the sources. Therefore, the characteristics of pollution sources include location of pollution sources, quantity, mode, intensity and time of emissions, protective measures, and type, toxicity, volatility, solubility, mobility and degradability of typical pollutants. The specific process of assessment is described below.

5.2 Pollution Load Assessment

5.2.1 Selection of Typical Pollutants

The typical pollutants are selected according to the characteristics of groundwater quality and pollution sources identified by the investigation conducted in the normal season of 2005, as shown in Table 5.2. They represent three integrated indicators, four inorganic matter indicators, three heavy metal indicators, and four organic matter indicators.

Table 5.2 Basic information of typical pollutants *Source* The table is reprinted from (Huan et al. 2018), with permission from Elsevier

Category	Pollutant	Source	Drinking water quality standards (mg/L)	lgK$_{oc}$	T$_{50}$(d)
Integrated indicators	TDS	Industry, households	1,000		
	COD	Industry, sewage rivers	5		
	TH	Industry, agriculture, households, and sewage rivers	450		
Inorganic matter indicators	SO$_4^{2-}$	Households, sewage rivers	250		
	Fluoride	Industry, sewage rivers	1		
	NH$_4$–N	Industry, agriculture, sewage rivers	0.5		
	NO$_3$–N	Industry, agriculture, households	20		
Heavy metal indicators	Hg	Industry	0.001		
	Cr	Industry	0.05		
	Zn	Industry	1.00		
Organic matter indicators	BaP	Industry, agriculture	0.00001	5.95	157.87
	Carbon tetrachloride	Agriculture	0.002	1.69	130.94
	Benzene	Industry	0.01	2.22	80.35
	Volatile phenol (phenol)	Industry	0.001	2.43	30.10

5.2.2 Load of Typical Pollutants

The load (*L*) of typical pollutants is measured by quantifying their mobility, toxicity and degradability, written as Formula (5.2).

$$L_{ij} = T_{ij}W_r + M_{ij}W_M + D_{ij}W_D \tag{5.2}$$

Wherein T_{ij}, M_{ij} and D_{ij} indicates the quantified values of toxicity, mobility and degradability of pollutant *i* respectively; W_T, W_M and W_D are the weights of toxicity, mobility and degradability respectively; and L_{ij} means the load of pollutant *i* from source *j* (no fundamental unit).

(1) Attribute quantification

For a certain number of typical pollutants, stronger toxicity implies greater destructive power and weaker environmental self-remediation capability (Vahidnia et al. 2007). The toxicity of typical pollutants can be measured with reference to the concentration limits specified in the *Drinking Water Quality Standards (2006)* (Table 5.2). A higher limit means weaker toxicity and smaller hazard.

The indicators for quantifying the mobility of organic pollutants and inorganic pollutants are different. In general, inorganic components are more mobile than organic components, which also varies depending on the groundwater environment. As to inorganic pollutants, the mobility in the groundwater is subject to water-rock interactions, including ion exchange, lyophilization, evaporation and concentration, mixture, and redox reactions. Considering the migration coefficients of chemical elements in surface water environment (Luo and Jin 1985), the typical inorganic pollutants in descending order of mobility are TDS, NO_3^-, SO_4^{2-}, NH_4^+, TH, and F^-. In regard to heavy metals, anions bonded with Cr^{6+} can migrate in many aqueous environments in oxidation state. The migration is strong in basic—slightly acidic conditions although limited by adsorption under acidic conditions (Bai and Wang 2009). Compared with zinc compounds, mercury compounds exhibit strong covalency and mobility in a natural environment or among organisms because of strong volatility and liquidity. Hence, Cr^{6+} is most mobile, followed by Hg, and Zn is least mobile. The mobility of organic pollutants, affected by adsorption, is measured by the organic carbon-water partition coefficient (LgK_{oc}) which characterizes the absorption by solid-phase organic carbon. A greater lgK_{oc} value means that the pollutant is more difficult to migrate and less harmful.

The degradability of typical pollutants is quantified with reference to biodegradation in the groundwater environment. The indicator of degradability is T_{50} which refers to the time (d) needed to half the concentration of organic matter from the original level. A greater T_{50} value speaks for stronger degradability and weaker hazard. In this study, EPI Suite is used to calculate lgK_{oc} and T_{50}, with results shown in Table 5.2. As an indicator to measure the content of organic matter

5.2 Pollution Load Assessment

in water, COD is considered most degradable. In regard to groundwater, NH_4^+ and NO_3^- are most readily biodegradable among inorganic pollutants, while TDS and the common inorganic contaminants are unlikely to degrade, reading SO_4^{2-}, F^- and TDS in descending order.

(2) Attribute weight

The groundwater pollution risk assessment classifies the attributes of typical pollutants as toxicity > mobility = degradability according to their importance in water supply safety and groundwater environmental self-remediation, and gives weights of 0.6, 0.2 and 0.2 respectively using the analytic hierarchy process (AHP).

(3) Load calculation

Normalization is needed due to dimensional non-uniformity when quantifying the toxicity, mobility and degradability of typical pollutants. As there are no clear quantitative reference values for mobility and degradability of inorganic and integrated pollutants, pollutant rankings by the three attributes are made, and the corresponding serial numbers represent the relative size of these attributes. A higher serial number speaks for greater harm of hazard. Where both reference value and serial number are the same, the remaining serial numbers are sequentially scaled. The results of pollutant load calculations are as shown in Table 5.3.

Table 5.3 Load of typical pollutants and the ranking *Source* The table is reprinted from (Huan et al. 2018), with permission from Elsevier

Pollutant	Toxicity ranking	Mobility ranking	Degradability ranking	L value	L ranking
COD	5	9	1	5	2
TDS	1	14	14	6.2	4
NH_4^+	8	11	2	7.4	7
NO_3^-	4	13	3	5.6	3
SO_4^{2-}	3	12	11	6.4	5
F	6	7	12	7.4	7
TH	2	8	10	4.8	1
Hg	12	6	8	10	12
Benzene	10	3	5	7.6	9
BaP	14	1	7	10	12
Carbon tetrachloride	11	4	6	8.6	11
Cr^{6+}	9	10	13	10	12
Volatile phenol	12	2	4	8.4	10
Zn	6	5	9	6.4	5

5.2.3 Calculation of Pollutant Emissions

(1) Industrial area

The industrial sources of pollution in the study area are distributed on both sides of the Songhua River as a variety of pollutants enter the water bodies through sewage outfalls and further pollute the groundwater. According to the *Report of Investigation to Groundwater Pollution in Key Sections of the Songhua River (Jilin City)* (2008), the 28 industrial point sources surveyed in 2005 are described, as shown in Table 5.4. Assuming emissions from these sources continue for two decades and using Darcy's law, the scope of outfall-specific influence on groundwater over the years is calculated based on the permeability coefficient and hydraulic gradient of aquifer. The scope is the buffer radius of these point sources.

In addition to discharge of industrial wastewater into the Songhua River, there is still leakage of contaminants at the industrial sites. It is assumed that 30% of the industrial wastewater leaks and 70% is discharged after centralized treatment.

Assuming that all industrial enterprises take measures to meet the *Integrated Wastewater Discharge Standards (1996)*, the discharge of pollutants at an industrial industrial zone is calculated as follows:

$$Q_{indus-pollu} = Q_{indus} \times R_1 \times \lambda_1 \tag{5.3}$$

Wherein Q_{indus} refers to the amount of industrial wastewater discharged in the industrial zone (t/a); $Q_{indus-pollu}$ is the amount of a specific pollutant discharged in the industrial zone (t/a); R_1 represents the pollutant emissions standards for the industrial zone (mg/L), which refers to the *Discharge Standards of Water Pollutants for Paper Industry* (GB 3544-2001) for paper mills; λ_1 is the infiltration rate, which is derived from the coefficient of precipitation infiltration with reference to *Jilin Hydrogeological Investigation Report* (2008).

(2) Agricultural area

There may be wastewater irrigation in the agricultural area, inferred from the existence of BaP and carbon tetrachloride in the groundwater. As the field survey did not differentiate clean water (including precipitation and groundwater), wastewater and reclaimed water for irrigation, it is assumed that dry fields, paddy fields and vegetable fields in the agricultural area are irrigated by clean water.

Urea, diammonium, special corn fertilizer, special rice fertilizer and compound fertilizer are used in the survey area, according to the *Report of Investigation to Groundwater Pollution in Key Sections of the Songhua River (Jilin City)* (2008). NH_4^+ in the fertilizers is partially volatilized and absorbed by plants in clean water irrigation area with a nitrogen use efficiency of 40% (Zhu 2000) and nitrogen loss rate of 36% (Su et al. 2006), and partially turned into nitrate nitrogen through soil nitrification. Nitrogen in nitrate is easy to absorb by plants and lose from surface runoff and subsurface leaching or through denitrification. It is assumed that over 85% of ammonia nitrogen in the soil is lost through vertical transportation and

5.2 Pollution Load Assessment

Table 5.4 Details on sewage outfalls

No.	Key enterprises	Sewage outfalls	Sewage discharge (10⁴ t/a) Whereabouts	Sewage discharge (10⁴ t/a) Post-treatment discharge	Major pollutants	Pollutant concentration (mg/L) COD	Ammonia	BaP	Benzene	Carbon tetrachloride	Zn	Cr^{6+}
1	JCGC Sewage Treatment Plant	JCGC Sewage Line	Songhua River	2,930	COD, ammonia nitrogen, BaP, benzene, carbon tetrachloride	90	25	3E-05	0.2	0.06	1	0.05
2	JCGC Acrylonitrile Plant	West Subline 10# (Acrylonitrile Plant Line 10#)	Songhua River	60	COD	150	50	3E-05	0.2	0.06	1	0.05
3	JCGC Fertilizer Plant	West Line 10#	Songhua River	1,027	COD,ammonia nitrogen	150	50	3E-05	0.2	0.06	1	0.05
4	JCGC Synthetic Resin Factory		Songhua River	16	COD,ammonia nitrogen	150	50	3E-05	0.2	0.06	1	0.05
5	Jilin Shuang'ou Chemical Co., Ltd.		Songhua River	3.8	COD,ammonia nitrogen	100	50	3E-05	0.2	0.06	1	0.05
6	Jilin Jiangbei Machinery Factory	Beidagou	Songhua River	8	COD,Zn,Cr^{6+}	150	50	3E-05	0.2	0.06	1	0.05
7	JCGC Pesticide Factory		Songhua River	2	COD, ammonia nitrogen, benzene	150	50	3E-05	0.2	0.06	1	0.05

(continued)

Table 5.4 (continued)

No.	Key enterprises	Sewage outfalls	Sewage discharge (10⁴ t/a)		Major pollutants	Pollutant concentration (mg/L)						
			Whereabouts	Post-treatment discharge		COD	Ammonia	BaP	Benzene	Carbon tetrachloride	Zn	Cr^{6+}
8	JCGC Diphenyl Plant	East Line 10#	Songhua River	450	COD, ammonia nitrogen, BaP, benzene, carbon tetrachloride	150	50	3E-05	0.2	0.06	1	0.05
9	JCGC Calcium Carbide Plant		Songhua River	821	COD	150	50	3E-05	0.2	0.06	1	0.05
10	Jilin Research Institute		Songhua River	3	COD, ammonia nitrogen	150	25	3E-05	0.2	0.06	1	0.05
11	JCGC Thermal Power Plant		Songhua River	3,066	COD	150	25	3E-05	0.2	0.06	1	0.05
9-1	JCGC Calcium Carbide Plant	South Line 10#	Songhua River	536	COD	100	25	3E-05	0.2	0.06	1	0.05
12	JCGC gyzstan Organic Synthesis Plant		Songhua River	6	COD	100	50	3E-05	0.2	0.06	1	0.05
13	JCGC Power Plant		Songhua River	173	COD	100	50	3E-05	0.2	0.06	1	0.05
14	JCGC Test Plant		Songhua River	10	COD,ammonia nitrogen	100	25	3E-05	0.2	0.06	1	0.05
15	Jilin Songtai Chemical Co., Ltd.		Songhua River	25	COD,ammonia nitrogen	100	50	3E-05	0.2	0.06	1	0.05

(continued)

5.2 Pollution Load Assessment

Table 5.4 (continued)

No.	Key enterprises	Sewage outfalls	Sewage discharge (10^4 t/a) Whereabouts	Post-treatment discharge	Major pollutants	Pollutant concentration (mg/L) COD	Ammonia	BaP	Benzene	Carbon tetrachloride	Zn	Cr^{6+}
16	JCGC Oil Refinery	Longtanchuan	Songhua River	24	COD	100	50	3E-05	0.2	0.06	1	0.05
17	JCGC Ethylene Plant		Songhua River	25	COD	100	50	3E-05	0.2	0.06	1	0.05
18	China Resources (Jilin) Brewery Co., Ltd.	China Resources Brewery Outfalls 1, 2	Songhua River	145	COD,ammonia nitrogen	150	25	3E-05	0.2	0.06	1	0.05
19	Jilin Jingwei Steel Manufacturing Co., Ltd		Songhua River	9	COD	100	50	3E-05	0.2	0.06	1	0.05
20	Jilin Jiuxin Chemical Plant	Jiuxin Chemical Plant 1	Songhua River	17.52	COD	146	50	3E-05	0.2	0.06	1	0.05
		Jiuxin Chemical Plant 2	Songhua River	17.52	COD	132	50	3E-05	0.2	0.06	1	0.05
		Jiuxin Chemical Plant 3	Songhua River	8.76	COD	160	50	3E-05	0.2	0.06	1	0.05

(continued)

Table 5.4 (continued)

No.	Key enterprises	Sewage outfalls	Sewage discharge (10^4 t/a) Whereabouts	Sewage discharge (10^4 t/a) Post-treatment discharge	Major pollutants	Pollutant concentration (mg/L) COD	Ammonia	BaP	Benzene	Carbon tetrachloride	Zn	Cr^{6+}
21	Jilin Chemical Fiber Co., Ltd	Jiuzhan Open Channel	Songhua River	2,500	COD	150	50	3E-05	0.2	0.06	1	0.05
22	Jilin Fuel Ethanol Co., Ltd	Fuel Ethanol	Songhua River	300	COD, ammonia nitrogen	150	25	3E-05	0.2	0.06	1	0.05
22-1	Jilin Chemical Fiber Co., Ltd	Chemical Fiber Co., Ltd 1, 2	Songhua River	1,940	COD, Zn	150	50	3E-05	0.2	0.06	1	0.05
23	Jilin Tuopai Agricultural Products Co., Ltd	Jilin Tuopai	Songhua River	7	COD, ammonia nitrogen	150	50	3E-05	0.2	0.06	1	0.05
24	Jilin Ferroalloy Co., Ltd	Ferroalloy Plant	Songhua River	850	COD	200	50	3E-05	0.2	0.06	1	0.05
25	Jilin Carbon Group Co., Ltd	Carbon Plant	Songhua River	387	COD	150	50	3E-05	0.2	0.06	1	0.05
26	Jidong Cement (Jilin) Co., Ltd	Jidong Cement	Songhua River	127	COD	150	50	3E-05	0.2	0.06	1	0.05
27	Jilin Chenming Paper Co., Ltd	Chenming Paper	Songhua River	280	COD, ammonia nitrogen	400	25	3E-05	0.2	0.06	1	0.05
28	Jilin Yuanyuan Thermal Power Co., Ltd		Songhua River	300	COD	70	25	3E-05	0.2	0.06	1	0.05

5.2 Pollution Load Assessment

surface runoff and the residual nitrate nitrogen enters the groundwater with precipitation (Cha et al. 2011).

The amount of nitrate nitrogen discharged to groundwater in agricultural area is calculated as follows:

$$Q_{agriculture} = (Q_{agri-total} \times S_{single}/S_{total}) \times \lambda_1 \quad (5.4)$$

Wherein $Q_{agriculture}$ stands for the quantity of nitrate nitrogen discharged in the agricultural area (t/a); $Q_{agri-total}$ stands for the quantity of fertilizer applied in the survey area, which is 12,663.4 t/a according to *Report of Investigation to Groundwater Pollution in Key Sections of the Songhua River (Jilin City)* (2008), S_{total} represents the total scope of fertilizer application, reading 227.27 km^2 and S_{single} individual agricultural area (km^2); λ_1 is the coefficient of precipitation infiltration without the fundamental unit, referring to *Jilin Hydrogeological Investigation Report* (2008).

(3) Rural residential area

There are a total of 19 rural settlements identified in the 2005 survey combined with land-use zoning map. Assuming that the domestic wastewater generated in the residential area is directly discharged on the surface without seepage control and migrates to the groundwater through infiltration, the discharge of pollutants in a single rural settlement is calculated as follows:

$$Q_{household} = Q_{sew-total} \times R \times \lambda \quad (5.5)$$

Wherein $Q_{household}$ stands for the amount of s a single pollutant discharged in the rural settlement into the groundwater (t/a); λ is the coefficient of precipitation infiltration without the fundamental unit; $Q_{sew-total}$ represents the quantity of domestic sewage discharge (m^3/a). In 2005, the discharge of domestic sewage was 22.35 tons per person in the study area (Yang 2013). The annual discharge of domestic sewage per unit of area amounted to 9,158.19 t/km^2·a in Longtan District (east of the Songhua River) and 17,353.82 t/km^2·a in Changyi District (west of the Songhua River), given 409.76 persons/km^2 and 776.46 persons/km^2 respectively according to *Social and Economic Statistical Yearbook of Jilin City* (2006). R indicates the concentration of typical pollutants (mg/L). The numbers are set to be 25.60, 5,965.70, 1.9, 2,086.25, 613 and 1,275 for nitrate nitrogen, COD, volatile phenol, ammonia nitrogen, TH and TDS respectively, based on landfill leachate concentrations in Longtan District in the *Report of Investigation to Groundwater Pollution in Key Sections of the Songhua River (Jilin City)* (2008) and pollutant concentrations of groundwater samples taken in the normal season in 2005.

(4) Polluted surface waters

According to the *Report of Investigation to Groundwater Pollution in Key Sections of the Songhua River (Jilin City)* (2008), Jiuzhan Open Channel is the polluted surface waters in the study area, which forms the linear source of pollution.

Assuming the open channels drain sewage for 20 years, the scope of pollutant diffusion in the vicinity is considered as the buffer radius according to Darcy's law. The amount of pollutant discharge per river is calculated as follows:

$$Q_{river} = L \times W \times V \times R \tag{5.6}$$

Wherein Q_{river} represents the amount of specific pollutant discharged from polluted rivers into groundwater (t/a). V is a permeability coefficient of sediment. The polluted waters result from the high fiber and lignin content of wastewater discharged prior to 1992 to the original natural rivers whose bed was sand and gravel. With reference to permeability coefficient of silt, the V value is set to be 2×10^{-9} m/s (Zheng and Gordon 2009). R indicates the concentration of typical pollutants (mg/L), and the values are taken from the *Report of Investigation to Groundwater Pollution in Key Sections of the Songhua River (Jilin City)* (2008). L stands for the channel length and W channel width, numbering 1,600 m and 5 m respectively. The calculation results are as shown in Table 5.5.

5.2.4 Pollution Load Evaluation Results

The potential hazard of pollution sources (S) is characterized by the load of typical pollutants (L) and the quantity of discharge (Q), calculated as in Formula (5.7).

$$S_j = \sum_{i=1}^{n} l_{ij} \times Q_{ij} \tag{5.7}$$

Wherein S_j represents the potential hazard of source j in the study area (no fundamental unit); Q_{ij} indicates the amount of pollutant i from source j (t/a); l_{ij} is the normalized value of L_{ij} (no fundamental unit).

Pollution load refers to the total quantity of pollutants released from sources per unit of time and per unit area. The sum of pollution loads from industrial, agricultural and household sources represents the hazard of all pollution sources per unit of area, which can be graded as shown in Fig. 5.2. Based on ArcGIS Union analysis, the comprehensive pollution load is divided into five levels using the Quantile method. A high level speaks for great harm of pollution. The comprehensive pollution load is mapped as shown in Fig. 5.3 and the statistical results are as shown in Table 5.6.

Table 5.5 Drainage of pollutants to groundwater from polluted surface waters

Category	Ammonia nitrogen	Sulfate	Fluoride	Volatile phenol	COD	TH
R(mg/L)	8.78	401	1.16	0.004	39.6	889
Q_{river}(t/a)	0.004534	0.207076	0.000599	2.07E-06	0.020449	0.45908

5.2 Pollution Load Assessment

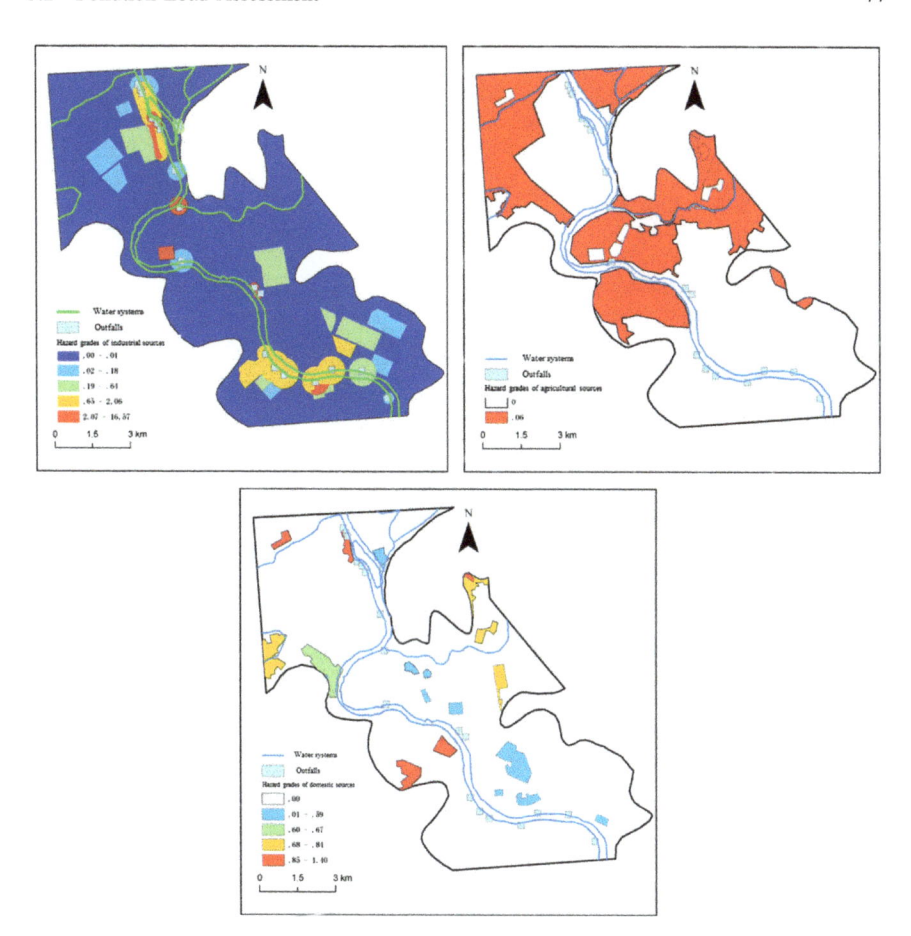

Fig. 5.2 Hazard grading of pollutants

According to Fig. 5.3 and Table 5.6, the pollution load is high in an area of 3.82 km² which accounts for 3.65% of the total study area, mainly in the vicinity of the outlets on the sides of the Songhua River, industrial sites and Jiuzhan Open Channel. In descending order, the high pollution load area includes Jiuzhan Open Channel and Chemical Fiber Plant (14.074–17.156), West Line 10# and Beidagou outfalls (6.623), JCGC sewage plant outfalls (4.999–5.059), Jilin sewage treatment plant (2.854–2.914), outfalls of Chenming Paper and East Line 10# (2.057–2.933).

The pollution load is relatively high in an area of 8.50 km², equivalent to 8.14% of the total study area. It involves outfalls along the Songhua River (ferroalloy plant, carbon plant and cement plant (1.032–1.693)), some industrial sites, residential communities and area in the vicinity of Jiuzhan Open Channel. These industrial sites include the diphenyl plant (1.954), thermal power plant (1.755) and pesticide plant (1.507), and the residential communities are located in the north bank of the Mangniu River (0.741–1.463), Qijiazi County (1.291–1.351), Jiuzhan

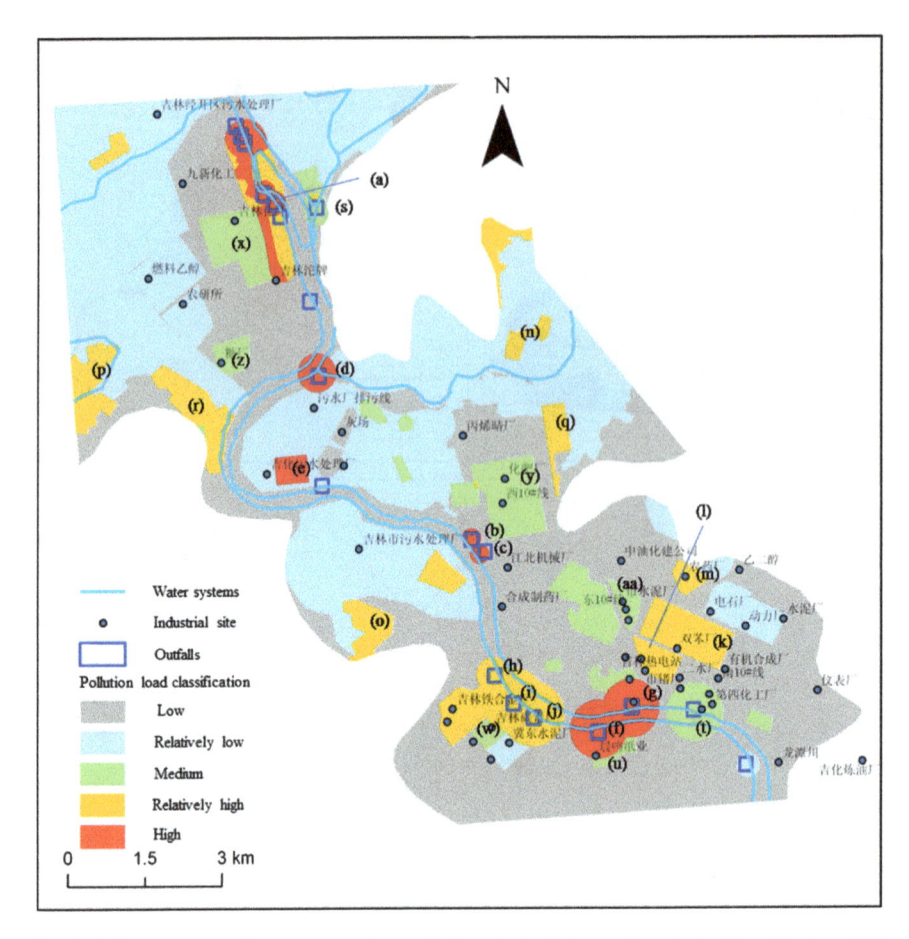

Fig. 5.3 Zoning map of comprehensive pollution load. Note: **a** Chemical Fiber Plant outfall; **b** West Line 10# outfall; **c** Beidagou outfall; **d** JCGC sewage plant outfall; **e** Jilin sewage treatment plant; **f** Chenming Paper outfall; **g** East Line 10# outfall; **h** Ferroalloy plant outfall; **i** Carbon plant outfall; **j** Cement plant outfall; **k** Diphenyl plant; **l** Thermal power plant; **m** Pesticide plant (**n**) North bank of the Mangniu River; **o** Qijiazi County; **p** Jiuzhan second-order terrace; **q** Qipan Street; **r** Jiuzhan Industrial Zone; **s** Huarun Brewery outfall; **t** South Line 10# outfall; **u** Chenming Paper Co., Ltd; **v** Huarun Brewery; **w** Jilin Carbon Plant; **x** Jilin Chemical Fiber Plant (0.531); **y** Fertilizer Plant (0.365); **z** Sugar Refinery (0.31); **aa** Municipal Cement Plant (0.235). The Figure is reprinted from (Huan et al. 2018), with permission from Elsevier

Table 5.6 Zoning results of comprehensive pollution load

Zone	Low	Relatively low	Medium	Relatively high	High
Pollution load range	<0.055	0.055–0.198	0.198–0.652	0.652–1.960	1.960–17.156
Area(km^2)	44.35	40.21	7.55	8.50	3.82
Percentage of area (%)	42.47	38.51	7.23	8.14	3.65

second-order terrace (0.842–1.122), Qipan Street on the south bank of the Mangniu River (0.741–0.801) and Jiuzhan Industrial Zone (0.674).

An area of 7.55 km^2 or 7.23% of the total study area is found with a medium level of pollution load. It covers residential communities, industrial sites, and some outfalls (Huarun beer plant (0.365), South Line 10# (0.340)), as well as agricultural areas (Qijiazi County, Jiangbei District (West), Mangniu River, Jinzhu District, Jiuzhan first-order terrace, 0.06). The residential communities are mainly located in Bajiazi Village of Jiangbei District, Jiangbei District (0.592) and the industrial sites include Chenming Paper Co., Ltd (0.637), Huarun Brewery (0.588), Jilin Carbon Plant (0.539), Jilin Chemical Fiber Plant (0.531), Fertilizer Plant (0.365), Sugar Refinery (0.31), and Municipal Cement Plant (0.235).

The area with low and relatively low pollution load, reading 84.56 km^2, accounts for 80.98% of the study area, mainly distributed in Jinzhu District, Jiuzhan first-order terrace, Mangniu River District, Jiangbei District, Hadawan District and Old Urban District.

In general, the pollution load is low or relatively low in a large percentage (80.98%) of the study area. About 15.37% of the study area suffers high or relatively high pollution load, primarily affected by sewage outfalls along the Songhua River and industrial and residential sources of pollution. The area of moderate pollution load (7.23%) is scattered in residential communities, industrial sites, sewage outfalls and agricultural fields (dry fields, paddy fields and vegetable fields).

5.3 Groundwater Function Evaluation

Groundwater function refers to the role and effect of the quality and quantity of groundwater systems of different scales and their spatial and temporal changes on the human society and the environment, covering groundwater resources, ecological environment, and geological environment (Zhang et al. 2009). Groundwater function evaluation, which reflects the expected hazard of the groundwater pollution system, is a risk assessment that characterizes the acceptable level of receptors.

In this study, the groundwater function evaluation follows the *Technical Specifications for Assessing and Zoning Groundwater Functions* (GWI-D5, 2006 Edition) printed and distributed by the Ministry of Land and Resources (hereinafter referred to as the Technical Specifications). The Technical Specifications are mainly applicable to the Quaternary groundwater system in the plains of the northwestern, northern and northeastern regions in China.

5.3.1 Indicator System and Rating

The indicators in the Technical Specifications are too complicated, and the data are regional and difficult to access. Based on these indicators, an indicator system

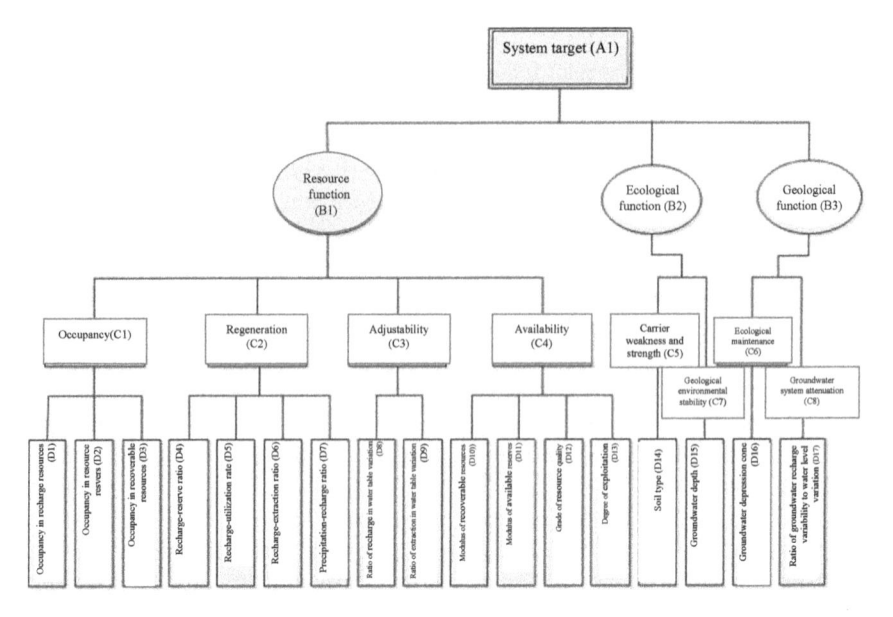

Fig. 5.4 Groundwater function evaluation system Source: The table is reprinted from (Huan et al. 2018), with permission from Elsevier

suitable for groundwater function evaluation of the study area is built, which takes into account basic data completeness and accuracy and hydrogeological characteristics, as shown in Fig. 5.4. Each indicator is described as follows:

The indicators need to be normalized to generally range from 0 to 1. In Table 5.2 of the Technical Specifications, indicators are grouped into three levels: strong (1), medium (0–1) and low (0), in order to represent the close correlation with groundwater. Since groundwater resources, ecological environment and geological environment are all related with groundwater, the extreme value normalization method is used to normalize the indicators within the range from 0.01 to 0.99, as shown in Formula (5.8).

$$Z_i = \frac{ax_i - \min(x_i)}{\max(x_i) - \min(x_i) + b} \tag{5.8}$$

Wherein a and b are coefficients for different indicators; *max* is the maximum value of an indicator of all units and *min* the minimum value; x_i is the actual value of unit i;

The various indicators of groundwater function are normalized below:

(1) Resource function (B1)

The resource function refers to the role or effect of groundwater supply and guarantee in certain conditions of supply, storage and renewal, with relatively independent and stable recharge sources and water supply capacity. It has four

5.3 Groundwater Function Evaluation

attributes: occupancy, regeneration, adjustability and availability. The indicators specific to attribute are described in Table 5.7.

The basic data required for indicators come from the *Report on Dynamic Groundwater Monitoring in Jilin Province* (1991–1995), *Report on Dynamic*

Table 5.7 Groundwater resource function indicators

Attribute layer (C)	Indicator layer (D)	
	Indicator	Description
Occupancy (C1)	Occupancy in recharge resources (D1)	Ratio of the modulus of recharge resources imported to evaluated zone or unit to to the average modulus of recharge resources in the study area
	Occupancy in resource reserves (D2)	Ratio of the modulus of groundwater resource reserves in the zone (belt) evaluated to the average modulus of groundwater resource reserves in the study area
	Occupancy in recoverable resources (D3)	Ratio of the modulus of available groundwater resources in the zone (belt) evaluated to the average modulus of available groundwater resource in the study area
Regeneration (C2)	Recharge-reserve ratio (D4)	Ratio of the modulus of recharge resources to the modulus of resource reserves in the zone (belt) evaluated
	Recharge-utilization rate (D5)	Ratio of the modulus of recharge resources to the modulus of available resources in the zone (belt) evaluated
	Recharge-extraction ratio (D6)	Ratio of annual average recharge to corresponding extraction in the zone (belt) evaluated in the recent 5–12 years
	Precipitation-recharge ratio (D7)	Ratio of annual average precipitation to corresponding recharge in the zone (belt) evaluated in the recent 5–12 years
Adjustability (C3)	Ratio of recharge in water table variation (D8)	Ratio of annual average recharge to corresponding water table variance in the zone (belt) evaluated in the recent 5–12 years
	Ratio of extraction in water table variation (D9)	Ratio of annual average extraction to water table variance in the zone (belt) evaluated in the recent 5–12 years
Availability (C4)	Modulus of recoverable resources (D10)	Amount of groundwater resources that can be mined per unit area of the zone (belt) evaluated
	Modulus of available reserves (D11)	Amount of groundwater resource reserves that can be used per unit area of the zone (belt) evaluated
	Grade of resource quality (D12)	Grade of groundwater quality in the zone (belt) evaluated, including I, II, III, IV and V
	Degree of exploitation (D13)	Ratio of annual average resources available to extracted in the zone (belt) evaluated in the recent 5–12 years

Groundwater Monitoring in Jilin Province (1996–2000), *Jilin Hydrogeological Investigation Report* (2008), *Report of Investigation to Groundwater Pollution in Key Sections of the Songhua River (Jilin City)* (2008), 84 drilling histograms, particle size analysis results of surface soil (2011) and three-year monitoring data of 16 groundwater table observation wells (2003–2005).

The spatial distribution of normalized resource function indicators is mapped, as shown in Fig. 5.5.

(2) Ecological function (B2)

The ecological function refers to the role or effect that the groundwater system has in the benign maintenance of terrestrial vegetation, lakes, wetlands or land quality.

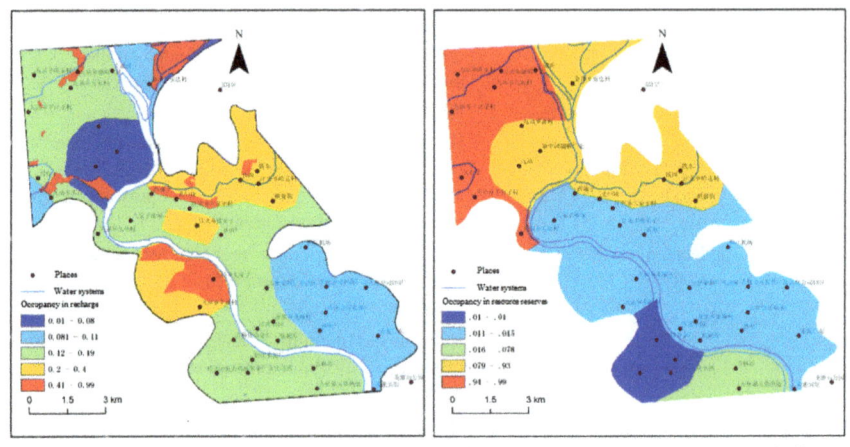

(a) Occupancy in recharge resources (D1)

(b) Occupancy in resource reserves (D2)

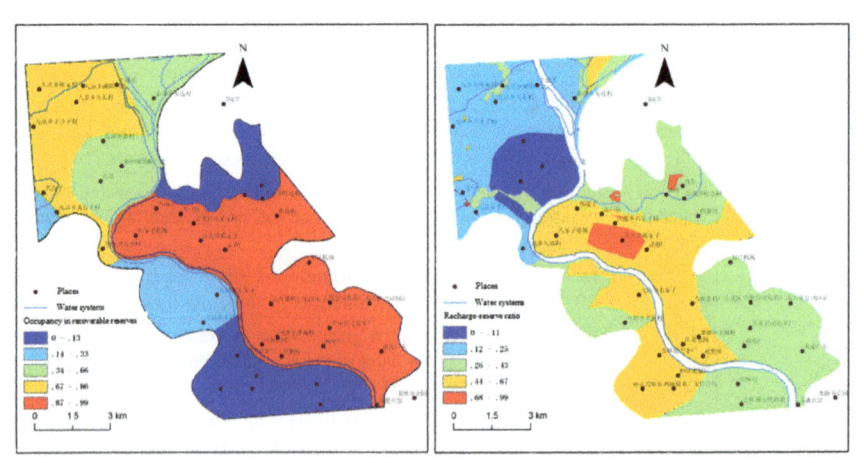

(c) Occupancy in recoverable resources (D3)

(d) Recharge-reserve ratio (D4)

Fig. 5.5 Grading maps of groundwater resource function indicators

5.3 Groundwater Function Evaluation

(e) Recharge-utilization rate (D5)

(f) Recharge-extraction ratio (D6)

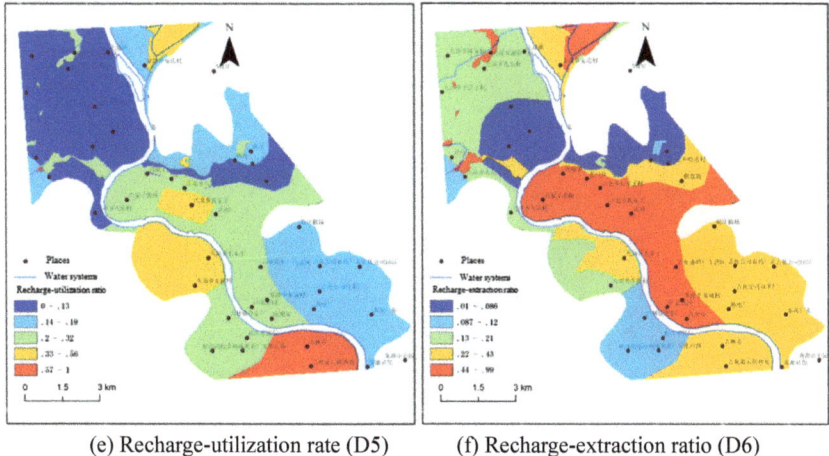

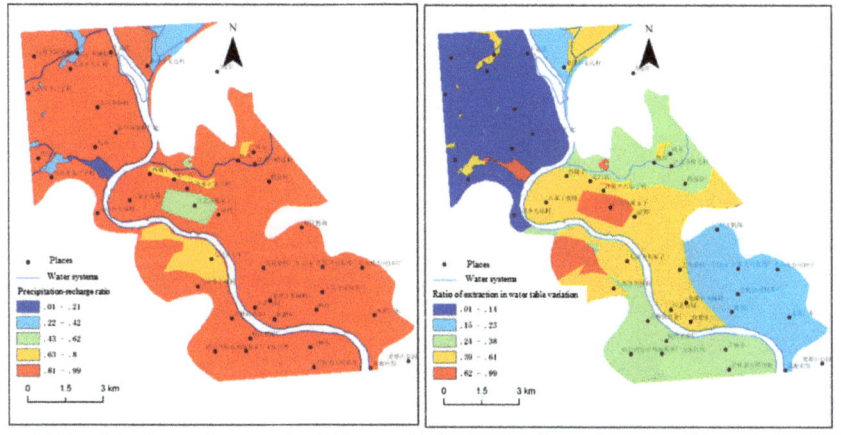

(g) Precipitation-recharge ratio (D7)

(h) Ratio of recharge in water table variation (D8)

Fig. 5.5 (continued)

In the Technical Specifications, the degree of groundwater correlation with ecological environmental factors, such as lake environment and landscape change index, is used as ecological function indicators. However, groundwater change within ecosystem carrying capacity has a relatively low intensity of stress on ecosystems. In this case, the degree of correlation is limited and the indicator solely is not enough to evaluate the ecological function. As there are no lakes and wetlands in the study area, the study uses vegetation to characterize the ecological environment and discusses the ecological effect of groundwater system on vegetation. Ecological function indicators include soil type and depth to water table, considering the local conditions.

(i) Ratio of extraction in water table variation (D9)

(j) Modulus of recoverable resources (D10)

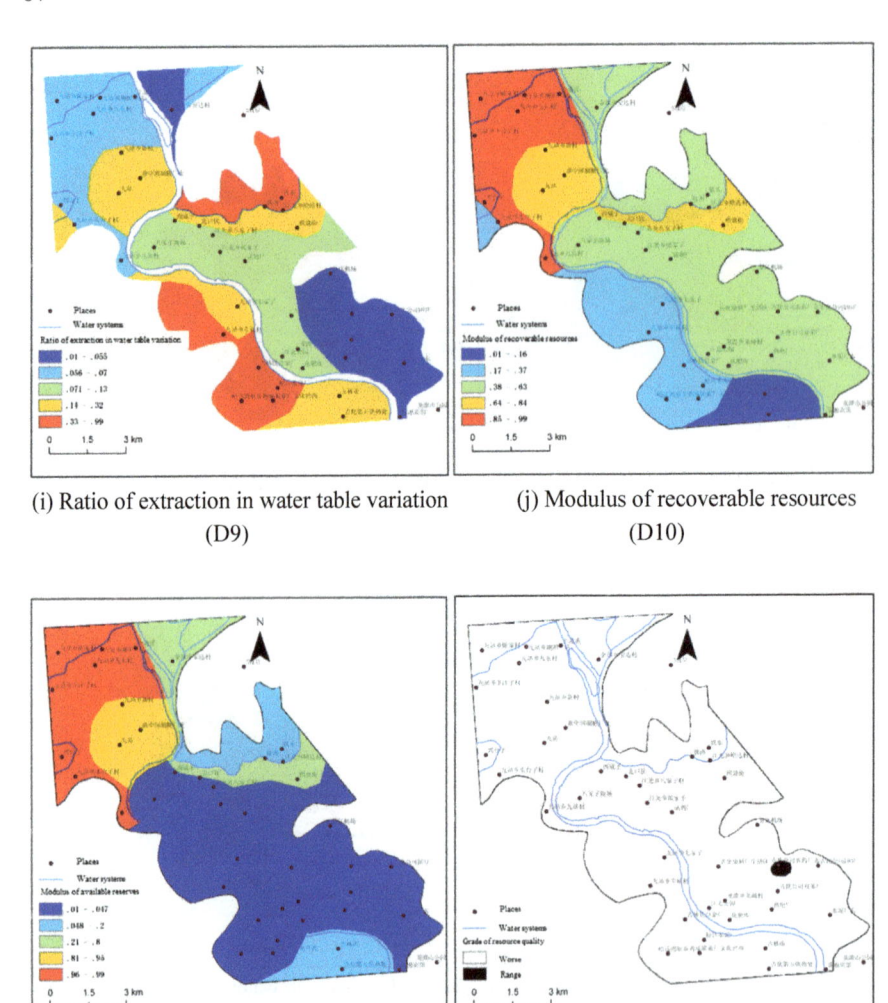

(k) Modulus of available reserves (D11)

(l) Grade of resource quality (D12)

Fig. 5.5 (continued)

A. Soil type. Soil particles and combinations of lithologic character in the vadose zone significantly affect vegetation growth and soil salinization. As the substrate of plant growth, soil provides water, minerals and other elements necessary to plant growth. The different properties of soils lead to different vegetation. According to the relevant ecological data, in terms of support for vegetation growth, soil particles of different lithologic character in descending order are clay, clay loam, sandy loam, sand and gravel. In the case of similar or consistent groundwater system conditions, the groundwater is easy to fulfill the functions in descending order of soils as

5.3 Groundwater Function Evaluation 85

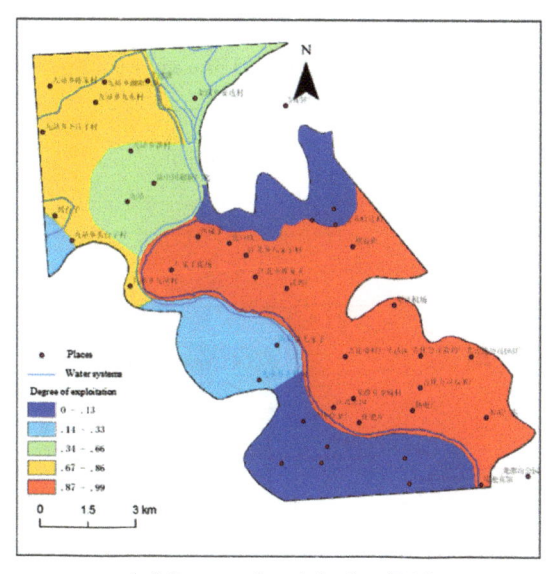

(m) Degree of exploitation (D13)

Fig. 5.5 (continued)

mentioned above. According to 28 soil samples taken 0–60 cm below ground and drilling histograms collected in 2011, the soils in the study area can be divided into loamy sand, sandy loam, sandy clay, silty loam, loam and silty clay, as well as gravel or cobble exposed to the surface.

B. Depth to water table. The water supply for vegetation growth is largely influenced by groundwater, in addition to precipitation and surface runoff. The vegetation on the surface reflect to a certain degree the quality of ecological environment in a region. The depth to water table is a key indicator of the presence of water and groundwater in the soil. Under normal circumstances, the larger depth to water table means less water content in the soil and worse quality of the ecological environment. According to the data collected at 34 monitoring sites of shallow groundwater table in the normal season in 2005, the depth to water table ranges from 2.5 to 10.2 m and averages 7.07 m, with a standard deviation of 1.89 m.

As shown in Table 5.8, the indicators of soil type and depth to water table are graded, with reference to ecological function evaluation results of groundwater in the Tarim Basin (Fan and Ma 2005), northwest arid and semiarid area and plain area of Jilin Province (Fan 2007). The indicators are given a value within the range [0.01, 0.99]. The distribution of indicator values is mapped, as shown in Fig. 5.6.

(3) Geological function (B3)

The geological function refers to the role or effect that the groundwater system supports and protects the stability of its geological environment. There are cones of depression in a small part of the study area due to excessive groundwater extraction,

Table 5.8 Grading of ecological function indicators

Soil type	Silty clay	Sandy clay/silty loam	Loam and sandy loam	Loamy sand	Gravel
Value	0.99	0.75	0.5	0.25	0.01
Depth to water table (m)	<3.0	3.0–4.0	4.0–6.0	6.0–8.0	>8.0
Value	0.25	0.99	0.6	0.25	0.01

(a) Soil type   (b) Depth to water table

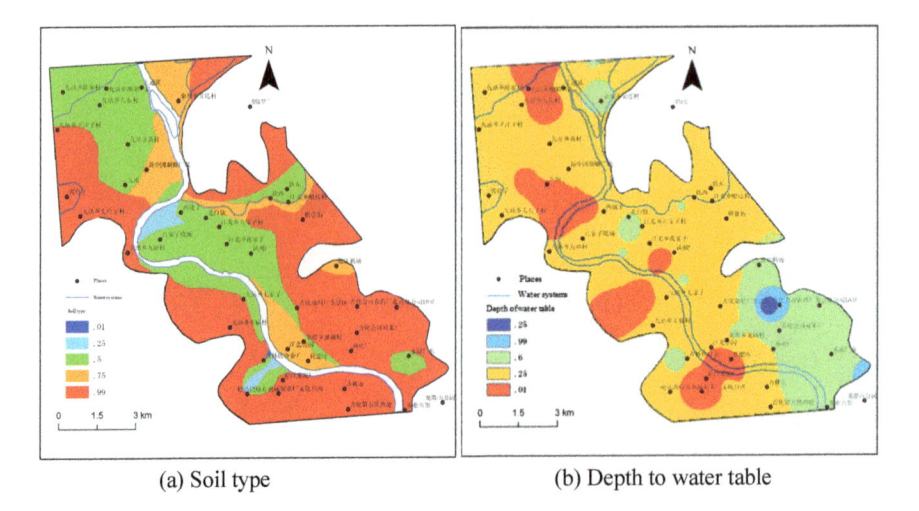

Fig. 5.6 Spatial distribution of ecological function indicators

which gradually threatens the stability of groundwater environment. Nevertheless, the geological environment is relatively stable in the study area as a whole and free from serious problems, such as large-scale salt water intrusion and land subsidence. The changes in the groundwater system did not cause significant deterioration or degradation of the geological environment. Given this, the cone of depression is used to characterize the stability of geological environment, and the ratio of groundwater recharge change rate to water table variation is used to characterize the attenuation of the groundwater system.

The cone of depression is a regional funnel-shaped concave resulted from drastic decline in groundwater tables associated with long-term groundwater overexploitation that makes exploitation quicker than replenishment and destroys the natural flow field. They pose a potential threat to the stability of regional geological environment. Herein, the cone of depression is a qualitative indicator to characterize the stability of geological environment in a region. Where there has been such a cone of depression, the demand index for geological function can be set to 0.99, indicating strong function and necessity to prohibit groundwater extraction and protect the geological environment. Otherwise, the demand index is 0.01, implying weak function of groundwater and tolerance for groundwater extraction.

5.3 Groundwater Function Evaluation

The ratio of groundwater recharge change rate to water table variation refers to the ratio of the rate of change in groundwater recharge to the water table variation during the the same period in the zone (belt) evaluated.

The geological function indicators are graded, as mapped in Fig. 5.7.

5.3.2 Indicator System Weight

The AHP method is used to identify the overall target and weight relationship of resource function (B1), ecological function (B2) and geological function (B3). In order to describe such target, the pairwise comparison of importance between Layer B indicators is conducted according to Layer A criteria and a matrix of the system (Layer A) can be obtained:

$$A = \{b_{ij} | i,j = 1 \sim n\} n \times m \tag{5.9}$$

Wherein b_{ij} represents the importance of b_i relative to b_j, that is B_i/B_j and $b_{i,j} = 1/b_{j,i}$.

On this basis, the comprehensive judging matrixes for weights of Layers A, B, C and D are constructed according to the characteristics of the study area. Then layers are ordered and weight vectors are solved, i.e., determining the ranking weight that shows the relative importance of factors in the same layer for a factor in the upper layer. These judging matrixes are tested and corrected to ensure consistency. The positive vector $W = (W_1, W_2, \ldots, W_n)^T$ is normalized as follows:

(a) Cone of depression

(b) Ratio of groundwater recharge change rate to water table variation

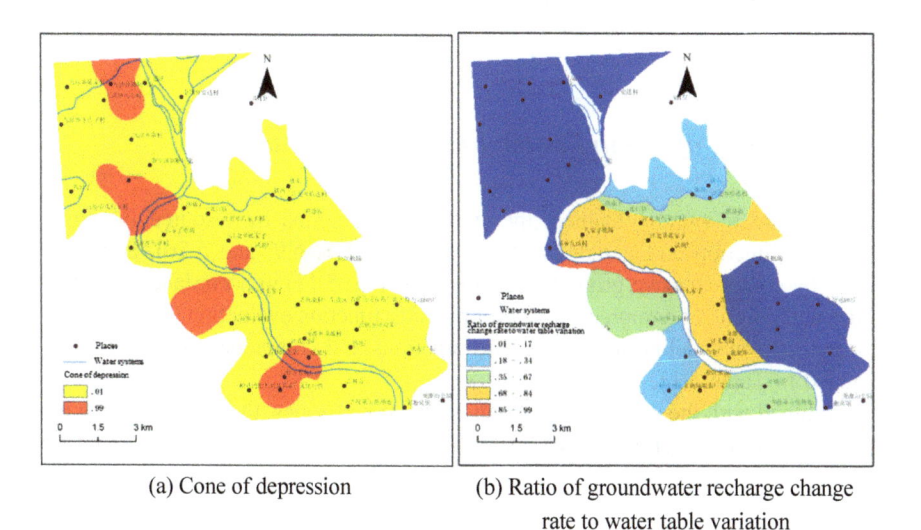

Fig. 5.7 Spatial distribution of geological function indicators

$$W_s = \left| \frac{W_1}{\sum_{i=1}^{n} W_i}, \frac{W_2}{\sum_{i=1}^{n} W_i}, \ldots, \frac{W_n}{\sum_{i=1}^{n} W_i} \right|^T \quad (5.10)$$

Where $W_s = (W_{s1}, W_{s2}, \ldots, W_{sn})^T$ represents the rating weight that shows the relative importance of factors in the same layer for a factor in the upper layer. In order to avoid the cumbersome calculation of data, the indicator system is scaled down according to the specific situation of the study area and the result of the weight ranking.

Further, the ranking weight of layers is determined. If Layer A includes m factors $B_1, B_2, \ldots, B_m$, the total ranking weight of Layer A is $a_1, a_2, \ldots, a_m$; if Layer B comprise n factors $C_1, C_2, \ldots, C_k, \ldots, C_n$, the ranking weight relative to factor B_j is $b_{1j}, b_{2j}, \ldots, b_{nj}$ (when C_k and B_j are irrelevant, $b_{kj} = 0$). After obtain the ranking weight of factors a_i, the comprehensive rating index (R) is calculated using the following formula:

$$R = \sum_{i=1}^{n} a_i X_i \quad (5.11)$$

Wherein a_i is the weight of the evaluation parameter and X_i represents the evaluation parameter, equal to $\sum (d_i/d_{\max})$ or $\sum (d_i/d_{阈})$, n is the number of evaluation parameters.

Through the establishment of a hierarchical structure (target layer, indicator level, strategy layer) using the AHP method, a judging matrix is formed, and based on hierarchical ranking calculation and consistency check, the weight of indicator is obtained, as shown in Table 5.9.

5.3.3 Results of Groundwater Function Evaluation

Based on the R values obtained, the study area is zoned by resource function, ecological function and geological function (Figs. 5.8, 5.9, and 5.10), and ultimately by comprehensive groundwater function (Fig. 5.11) in accordance with the grades and criteria of comprehensive groundwater function evaluation (Table 5.10). The area of zones delineated by specific functions and comprehensive function is provided in Table 5.11.

Combined with the spatial distribution of functional zones, the study area can be divided according to groundwater sustainability.

5.3 Groundwater Function Evaluation

Table 5.9 Indicator weights for groundwater function evaluation *Source* The table is reprinted from (Huan et al. 2018), with permission from Elsevier

System layer	Function layer B	Property layer C	Indicator layer D
Ground water function	Resource function 0.540	Occupancy 0.467	Occupancy in recharge resources (D1) 0.163
			Occupancy in resource reserves (D2) 0.297
			Occupancy in recoverable resources (D3) 0.540
		Regeneration 0.160	Recharge-reserve ratio (D4) 0.122
			Recharge-utilization rate (D5) 0.227
			Recharge-extraction ratio (D6) 0.423
			Precipitation-recharge ratio (D7) 0.227
		Adjustability 0.095	Ratio of recharge in water table variation (D8) 0.4
			Ratio of extraction in water table variation (D9) 0.6
		Availability 0.278	Modulus of recoverable resources (D10) 0.351
			Modulus of available reserves (D11) 0.189
			Grade of resource quality (D12) 0.351
			Degree of exploitation (D13) 0.108
	Ecological function 0.297	Carrier weakness and strength (0.3)	Soil type (D14) 1.0
		Ecological maintenance (0.7)	Depth to water table (D15) 1.0
	Geological function 0.163	Geological environmental stability (0.5)	Cone of depression (D16) 1.0
		Groundwater system attenuation (0.5)	Ratio of groundwater recharge change rate to water table variation (D17) 1.0

Strong sustainability: 12.88% of the study area or 12.85 km^2 shows strong groundwater sustainability, mainly distributed in the first-order terrace of Jiuzhan District and south bank of the Mangniu River. Groundwater in this zone has great resource function, followed by ecological function and geological function. It provides effective support for the sound development of the ecological and geological environment. While large-scale extraction is allowed, the ecological function should be protected and the geological function may be weakened.

(a) Occupancy

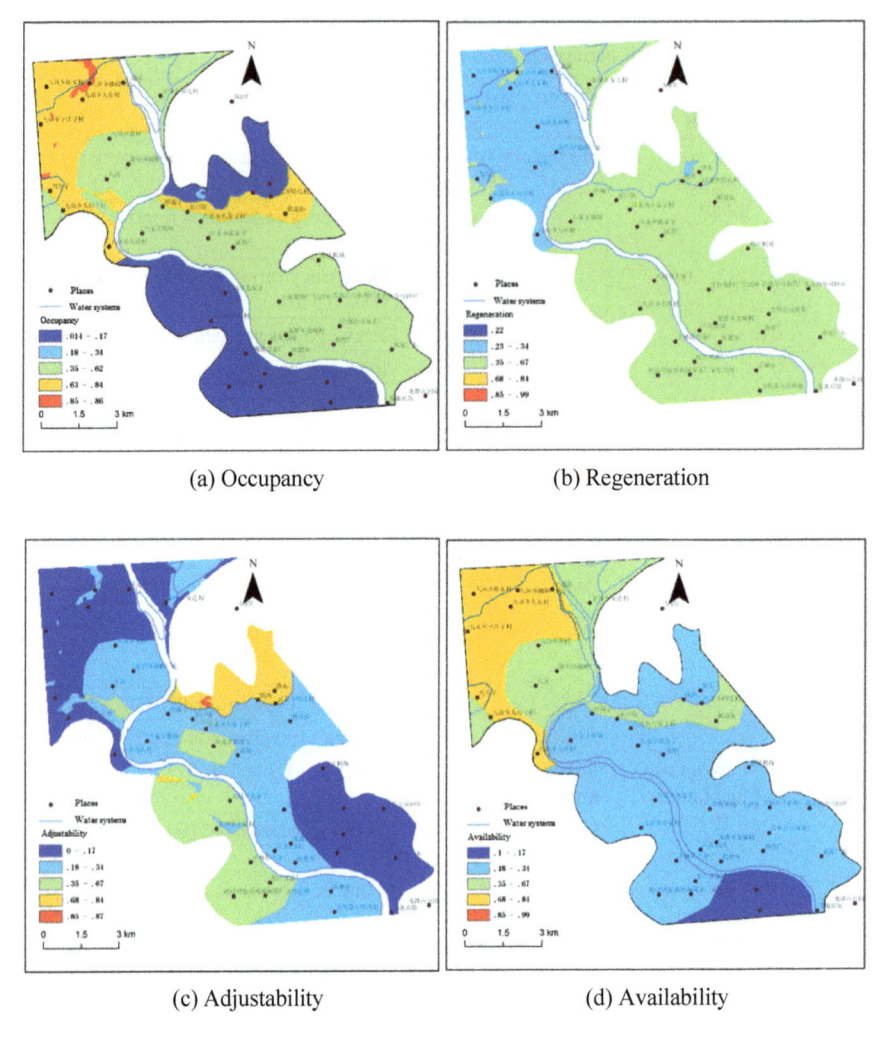

(b) Regeneration

(c) Adjustability

(d) Availability

Fig. 5.8 Groundwater function evaluation results

Relatively strong sustainability: 40.95% of the study area or 40.87 km² shows relatively strong groundwater sustainability, mainly distributed in Jiuzhan Industrial Zone, first-order terrace of Jiuzhan and Jinzhu Districts, south bank of the Mangniu River and most areas of Jiangbei District. Groundwater as a resource functions moderately or even better in this zone. The ecological function of groundwater is more complex, ranging from strong to weak, while the geological function is either moderate or weak. Overall, the resource function and ecological function overweigh than the geological function.

5.3 Groundwater Function Evaluation

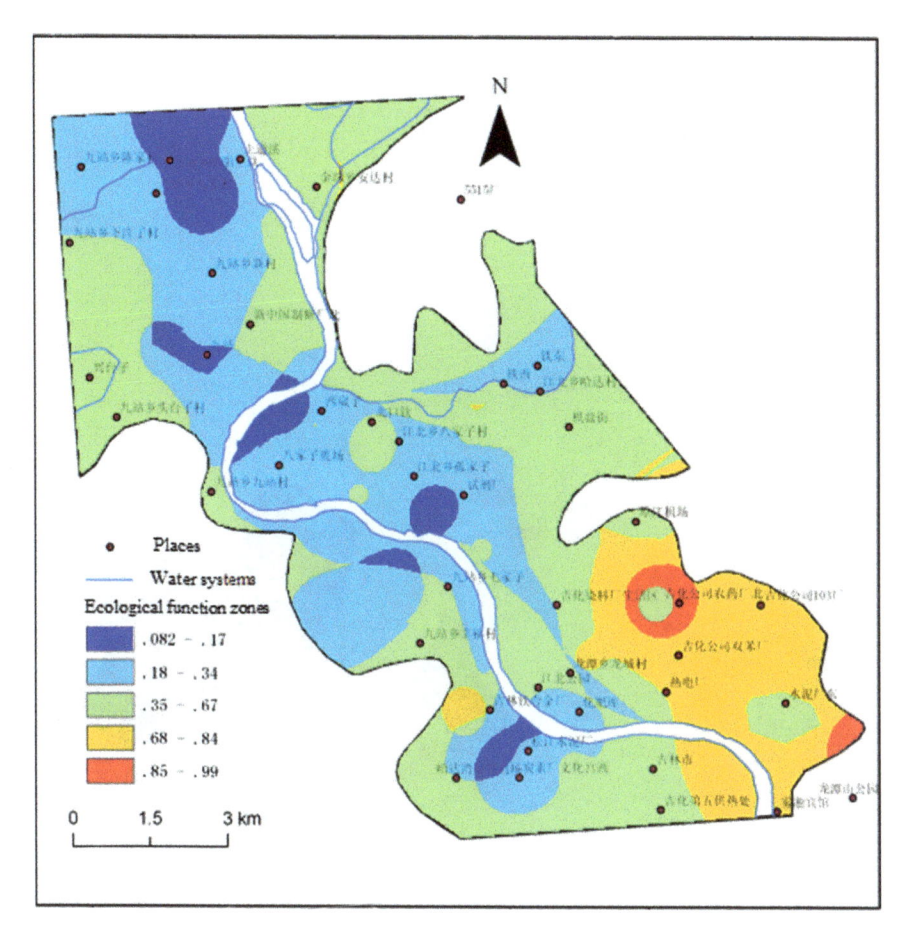

Fig. 5.9 Ecological function evaluation results

Moderate sustainability: A medium level of groundwater sustainability is found in Jiuzhan Industrial Zone, second-order terrace of Jiuzhan and Jinzhu Districts, and western area of Jiangbei District, which covers an area of 18.05 km^2 and represents 18.09% of the study area. Groundwater has general resource function and general or weaker ecological function and geological function. Yet, the resource function is overshadowed by the ecological function, while the geological function is the strongest among the three. Hence, appropriate groundwater extraction is allowed and the functions concerning the ecological and geological environment may be weakened.

Relatively weak sustainability: 17.19% of the study area or 17.16 km^2 shows relatively weak sustainability, mainly distributed in the north bank of the Mangniu River, Qijiazi County, most area of Hadawan District, and some area of the Old Urban District. Groundwater has relatively weak resource function or weak in the eastern high plains, and ecological and geological functions of different levels in

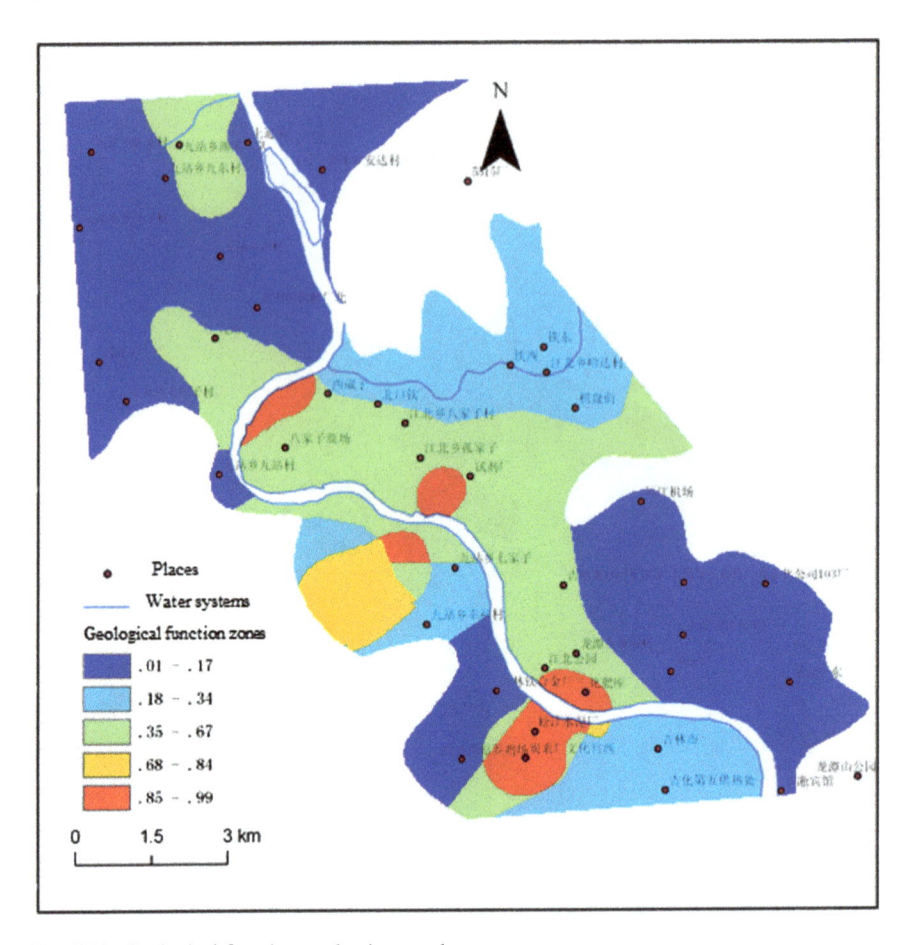

Fig. 5.10 Geological function evaluation results

different areas. On the whole, the ecological and geological functions overweigh the resource function and the groundwater sustainability is weak.

Week sustainability: 10.89% of the study area or 10.87 km² shows weak sustainability, mainly distributed in the south bank of the Mangniu River, Hadawan District, and some area of the Old Urban District. The groundwater function is assessed to be weak due to weak resource function and relatively weak ecological and geological functions. Overall, the resource function is overshadowed by ecological and geological functions and the groundwater sustainability hereof is worst in the study area.

5.3 Groundwater Function Evaluation

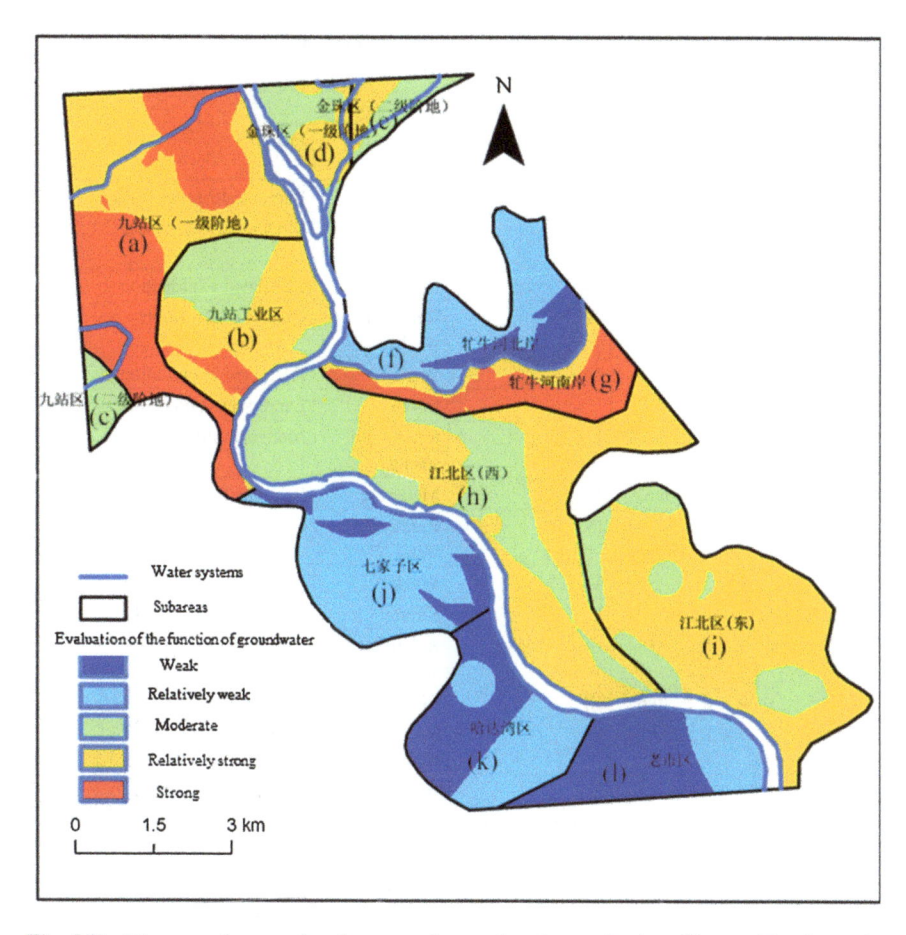

Fig. 5.11 Diagram of comprehensive groundwater function evaluation. Note: **a** The first-order terrace of Jiuzhan District; **b** Jiuzhan Industrial Zone; **c** The second-order terrace of Jiuzhan District; **d** The first-order terrace of Jinzhu District; **e** The second-order terrace of Jinzhu District; **f** North bank of Mangniu River; **g** South bank of Mangniu River; **h** Jiangbei District (West); **i** Jiangbei District (East); **j** Qijiazi County; **k** Hadawan District; **l** Old Urban District. The Figure is reprinted from (Huan et al. 2018), with permission from Elsevier

Table 5.10 Ratings and criteria for groundwater function evaluation *Source* The table is reprinted from (Huan et al. 2018), with permission from Elsevier

(a) Ratings and significance for individual groundwater function

Functions and codes	Rating (R)	State level	Functional status	Prospect
Resource Function (B1)	R > 0.84	I	Strong	Large-scale exploitation
	0.67 < R ≤ 0.84	II	Relatively strong	Moderate exploitation
	0.34 < R ≤ 0.67	III	Moderate	Regulated exploitation
	0.17 < R ≤ 0.34	IV	Relatively weak	Careful exploitation
	R ≤ 0.17	V	Weak	Restricted exploitation
Ecological function (B2)	R > 0.84	I	Strong	Large-scale utilization
	0.67 < R ≤ 0.84	II	Relatively strong	Moderate utilization
	0.34 < R ≤ 0.67	III	Moderate	Regulated utilization
	0.17 < R ≤ 0.34	IV	Relatively weak	Careful utilization
	R ≤ 0.17	V	Weak	Restricted utilization
Geological function (B3)	R > 0.84	I	Strong	Protection
	0.67 < R ≤ 0.84	II	Relatively strong	Conservation
	0.34 < R ≤ 0.67	III	Moderate	Utilization
	0.17 < R ≤ 0.34	IV	Relatively weak	Dilution
	R ≤ 0.17	V	Weak	Weakening

(b) Ratings and criteria for comprehensive groundwater function

Rating	0–0.2	0.2–0.4	0.4–0.6	0.6–0.8	0.8–1.0
Criteria	Weak groundwater sustainability	Relatively weak groundwater sustainability is weak	Moderate groundwater sustainability	Relatively strong groundwater sustainability	Strong groundwater sustainability

Table 5.11 Percentage of area by groundwater function grades (%)

Functional components	Function grading				
	Strong	Relatively strong	Moderate	Relatively weak	Weak
Resource function	17.30	15.15	39.38	0.34	27.83
Ecological function	1.49	12.12	46.01	34.07	6.31
Geological function	4.58	3.07	28.47	20.10	43.79
Comprehensive function	12.88	40.95	18.09	17.19	10.89

5.4 Results and Verification

5.4.1 Results

Based on the spatial distribution, intrinsic groundwater vulnerability, pollution load and groundwater function are integrated to form the results of groundwater pollution risk assessment using the ArcGIS-based matrix method. More specifically, the urgency of groundwater protection is mapped based on intrinsic groundwater vulnerability and groundwater function to represent groundwater pollution consequences, and the pollution load map is used to represent groundwater pollution possibility. The two maps are combined to form the matrix for groundwater pollution risk assessment (Table 5.12).

Figure 5.12 shows the zoning map for groundwater pollution risk in the study area using index overlay method. Table 5.13 provides statistics about zones of different shallow groundwater pollution risk in the study area.

As shown in Fig. 5.12, the high-risk area, covering 2.27 km², accounts for only 2.18% of the study area, mainly in the vicinity of outfalls on the west side of the

Table 5.12 Groundwater pollution risk assessment matrix

Urgency of groundwater protection		Intrinsic groundwater vulnerability				
		Low	Relatively low	Moderate	Relatively high	High
Groundwater functionvalue	High	Moderate	Relatively high	Relatively high	High	High
	Relatively high	Relatively low	Moderate	Relatively high	Relatively high	High
	Moderate	Relatively low	Relatively low	Moderate	Relatively high	Relatively high
	Relatively low	Low	Relatively low	Relatively low	Moderate	Relatively high
	Low	Low	Low	Relatively low	Relatively low	Moderate
Groundwater pollution risk		Urgency of groundwater protection				
		Low	Relatively low	Moderate	Relatively high	High
Pollution load	High	Moderate	Relatively high	Relatively high	High	High
	Relatively high	Relatively low	Moderate	Relatively high	Relatively high	High
	Moderate	Relatively low	Relatively low	Moderate	Relatively high	Relatively high
	Relatively low	Low	Relatively low	Relatively low	Moderate	Relatively high
	Low	Low	Low	Relatively low	Relatively low	Moderate

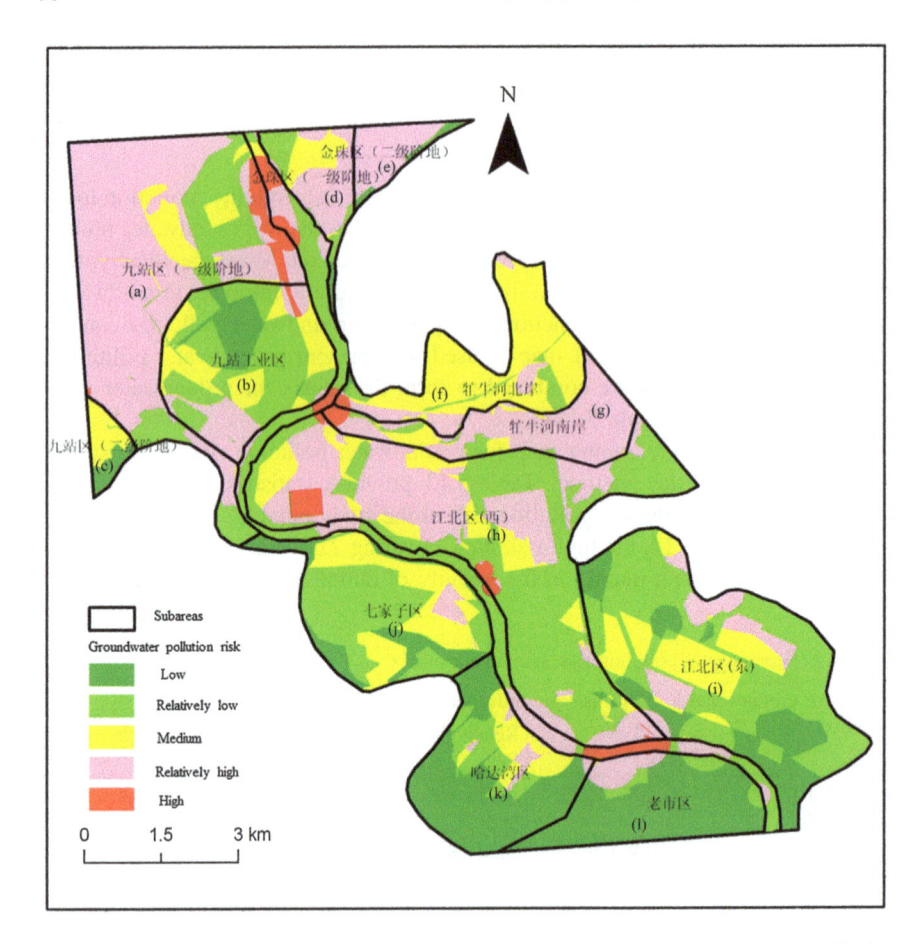

Fig. 5.12 Groundwater pollution risk mapping. Note: **a** The first-order terrace of Jiuzhan District; **b** Jiuzhan Industrial Zone; **c** The second-order terrace of Jiuzhan District; **d** The first-order terrace of Jinzhu District; **e** The second-order terrace of Jinzhu District; **f** North bank of Mangniu River; **g** South bank of Mangniu River; **h** Jiangbei District (West); **i** Jiangbei District (East); **j** Qijiazi County; **k** Hadawan District; **l** Old Urban District. The Figure is reprinted from (Huan et al. 2018), with permission from Elsevier

Table 5.13 Grading shallow groundwater pollution risk using the index overlay method

Groundwater pollution risk	Low	Relatively low	Medium	Relatively high	High
Area (km^2)	16.26	35.17	19.02	31.64	2.27
Percentage of the total area (%)	15.58	33.70	18.22	30.31	2.18

Mangniu River and along the Songhua River, such as Jiuzhan Open Channel, Jiuzhan Chemical Plant, and Chemical Fiber Factory. The pollution is mainly caused by industrial point sources and linear sources.

The groundwater pollution risk is relatively high in an area of 31.64 km^2, equivalent to 30.31% of the study area. Such area covers the floodplains of the Songhua River and the Mangniu River, first-order terraces (including first-order terrace of Jiuzhan District, Jinzhu District, south bank of the Mangniu River, northern (northwestern) area of Jiangbei District) and second-order terraces (in the vicinity of pesticide plant, thermal power plant, calcium carbide plants, and cement plant in Jiangbei District (east)), as well as surrounding areas of outfalls. The pollution is mainly caused by agricultural non-point sources (paddy fields, dry fields and vegetable fields) and in certain areas, industrial point sources.

The medium-risk area is 19.02 km^2, accounting for 18.22% of the study area and distributed mainly in the north bank of the Mangniu River, Jiangbei District (west), Qijiazi and Hadawan floodplains, and Jiangbei District (East), as well as sporadically in Jiuzhan District and Jinzhu District. The groundwater pollution is relatively low in an area of 35.17 km^2 or 33.70% of the study area, including the Jiuzhan Industrial Zone, Qijiazi County and Jiangbei District. The low-risk area is distributed in Hadawa District and the Old Urban District, which covers 16.26 km^2 or 15.28% of the study area.

5.4.2 Result Verification

The reliability of groundwater pollution risk assessment results is tested using the level difference method. The principle is to rate groundwater pollution and groundwater pollution based on the evaluation results. According to Sect. 4.3, groundwater pollution is classified into five levels: no, slight, moderate, heavy and severe, which is rated 1, 2, 3, 4 and 5 respectively; groundwater pollution risk is also divided into five levels: low, relatively low, medium, relatively high, high, which is rated 1, 2, 3, 4 and 5 respectively. The absolute value of level difference between the two dimensions is calculated for each unit of the study area. When the absolute value is between 0 and 1, the results of groundwater pollution risk assessment are deemed reasonable. When the risk level is 2 or 3 higher than the pollution level, the groundwater pollution risk is overestimated. When such absolute level difference is larger than 4, the groundwater pollution risk is completely overestimated (Stigter et al. 2006).

The level difference between groundwater pollution and groundwater pollution risk is mapped, as shown Fig. 5.13. According to statistical analysis, the reasonable area accounts for 64.45% of the study area, where there are 14 sampling sites, representing 70.8% of the total. It means that the risk assessment methodology is reasonable and applicable to the study area. As the pollution level is 2–3 higher than risk level, the groundwater pollution risk is underestimated in 35.5% of the study area, mainly in Jiuzhan Industrial Zone and Hadawan District, central Jiangbei

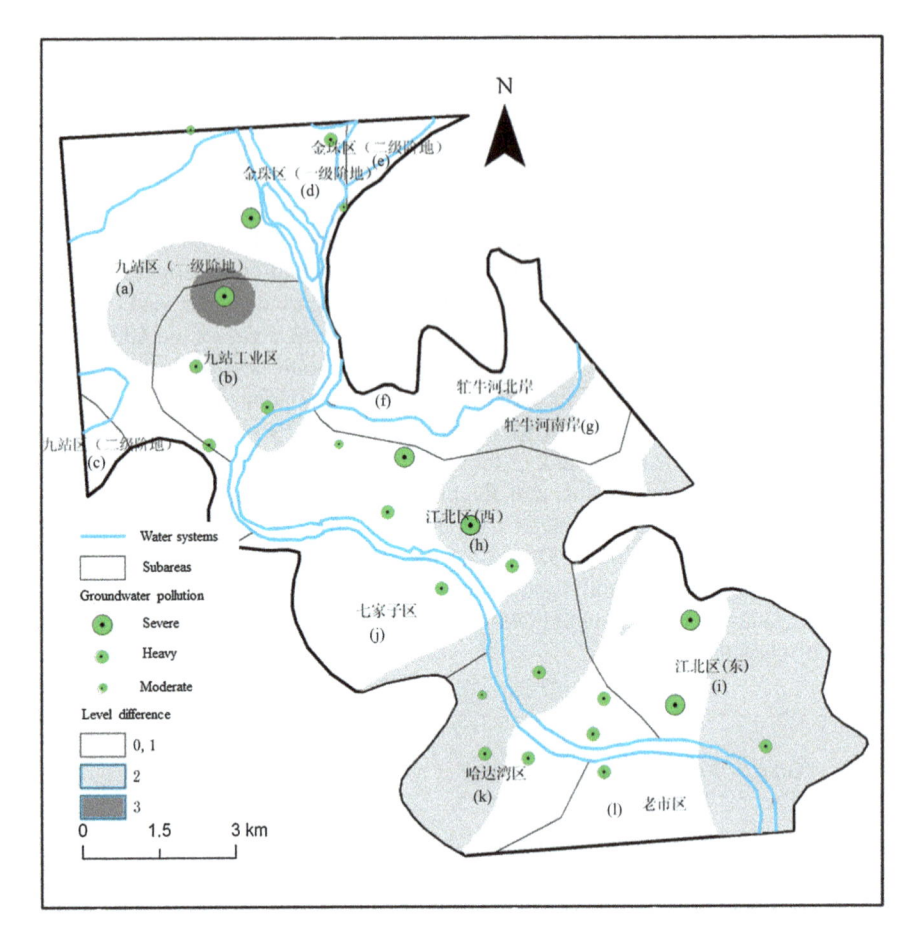

Fig. 5.13 Level difference between groundwater pollution and groundwater pollution risk. Note: **a** The first-order terrace of Jiuzhan District; **b** Jiuzhan Industrial Zone; **c** The second-order terrace of Jiuzhan District; **d** The first-order terrace of Jinzhu District; **e** The second-order terrace of Jinzhu District; **f** North bank of Mangniu River; **g** South bank of Mangniu River; **h** Jiangbei District (West); **i** Jiangbei District (East); **j** Qijiazi County; **k** Hadawan District; **l** Old Urban District. The Figure is reprinted from (Huan et al. 2018), with permission from Elsevier

District (west) and eastern Jiangbei District (east). The errors may be caused by two factors: (1) incomplete survey of pollution sources. The study only investigated industrial sewage outfalls and made assumptions on agricultural sources and domestic sources based on statistical yearbooks, relevant literature, and reports, so the accuracy of pollution load evaluation results needs to be improved. (2) The sampling sites are limited in the study area and mainly distributed in the Songhua River floodplain. Groundwater pollution zoning based on the samples may be inaccurate.

References

Aller L, Lehr JH, Petty R, et al. DRASTIC: a standardized system to evaluate ground water pollution potential using hydrugedlugic settings. Environment Protection Agency (EPA) (Number: 600287035); 1987.

Bai LP, Wang YY. A study of chromium migration and transformation in the soil and groundwater. Geol Resour. 2009;18(2):144–8.

Cha ZZ, Yan L, Hua YG. A preliminary study of vertical transportation characteristics of urea derived nitrate nitrogen in three types of latosols of parent materials. Chin J Trop Crops. 2011;32(5):821–7.

Editorial Board of Social and Economic Statistical Yearbook of Jilin City. Social and economic statistical yearbook of Jilin City. Beijing: China Statistics Press; 2006.

Fan W. Groundwater function evaluation of the plain area of Jilin Province. Changchun: Jilin University; 2007.

Fan ZL, Ma YJ. Ecological groundwater table and rational depth in the Tarim River Basin. Resour Sci. 2005;27(1):8–13.

Groundwater dynamic monitoring report in Jilin City of Jilin province (1991–1995). Environmental Monitoring Center of Jilin Province; 1997.

Groundwater dynamic monitoring report in Jilin City of Jilin province (1996–2000). Environmental Monitoring Center of Jilin Province; 2002.

Huan H, Wang JS, Zhai YZ, et al. Quantitative evaluation of specific vulnerability to nitrate for groundwater resource protection based on process-based simulation model. Sci Total Environ. 2016;550:768–84.

Huan H, Zhang BT, Kong HM, et al. Comprehensive assessment of groundwater pollution risk based on HVF model: a case study in Jilin City of northeast China. Sci Total Environ. 2018;628–629:1518–30.

Jilin Hydrogeological Investigation Report. Environmental Monitoring Center of Jilin Province; 2008.

Luo WY, Jin MQ. Preliminary exploration to the relationship between water migration coefficient and degree of karst development. China Karst. 1985;1(Z1):14–21.

Report of investigation to groundwater pollution in key sections of the Songhua River (Jilin City). Environmental Monitoring Center of Jilin Province; 2008.

Stigter TY, Ribeiro L, Carvalho DAMM. Evaluation of an intrinsic and a specific vulnerability assessment method in comparison on a regional scale. J Hydrogeol J. 2006;14(3):79–99.

Su F, Huang BX, Ding XQ, et al. Comparison of volatilization loss of nitrogen in different forms of ammonia fertilizer. Soil. 2006;38(6):682–6.

Vahidnia A, Van Der Voet GB, De Wolf FA. Arsenic neurotoxicity- a review. Hum Exp Toxicol. 2007;26(10):823–32.

Yang LH. A study of economic and environmental effects of Songhua River Basin (Jilin Province) and industrial space organization. Beijing: University of Chinese Academy of Sciences; 2013.

Zhang GH, Yang LZ, Nie ZL, et al. Characteristics and assessment of groundwater functions in North China Plain. Resour Sci. 2009;31(3):368–74.

Zheng CM, Gordon DB. Groundwater contaminant migration modeling. 2nd ed. Beijing: Higher Education Press; 2009.

Zhu ZL. Loss of nitrogen fertilizer in farmland and countermeasures. Soil Environ Sci. 2000;9(1):1–6.

Chapter 6
Economic Losses of Groundwater Pollution

Abstract Recognizing the value of environmental resource, the developed countries and regions around the world are reforming their national economic account system by including economic losses of environmental resources, so as to truly reflect the development of the national economy. The integration of economic losses caused by pollution-included groundwater depletion into economic construction costs is of high significance to environmental protection, resource conservation and national economic development. Based on the results of groundwater pollution risk assessment, this chapter estimates the economic losses caused by groundwater pollution are calculated, which intuitively provides a dynamic and scientific basis for groundwater environmental protection and governance. The results support the establishment and implementation of groundwater pollution prevention and control system and the sustainability of economic development.

Keywords Aggregate value of groundwater resources · Concentration-value curve Comprehensive loss rate · Total economic losses

6.1 Aggregate Value of Groundwater Resources

The aggregate value of groundwater resources encompasses the base value and the compensated value (Liu 2007).

$$k = P_d + P_b = (Q \times L - B) + (H_b + K_b + Z_b) \tag{6.1}$$

Wherein k is the total value of resources, P_d the base value and P_b the compensated value; L represents the market price of resources, Q the total output of resources and B the total costs for production and development of resources; H_b indicates the compensation for ecological damage, K_b compensation for geological prospecting and exploration, and Z_b compensation for resource development.

(1) Quantity of groundwater pollution (Q)

According to the results of groundwater pollution evaluation, there are no pollution-free zones and slightly polluted zones in the study area. Therefore, it is assumed that all the unconfined aquifers are contaminated. Groundwater pollution quantity ($Q_{pollution}$) is equal to the product of polluted aquifer volume (V) and storativity (μ):

$$Q_{pollution} = V_{pollutedaquifers} \times \mu = S_{pollutedaquifers} \times H_{aquifers} \times \mu \quad (6.2)$$

Wherein the value of μ is drawn from *Jilin Hydrogeological Investigation Report* (2008), and $H_{aquifers}$ represents the average height of the aqueous layers in different sections. The quantity of groundwater pollution is calculated to reach 3.0057×10^8 m^3, as shown in Table 6.1.

(2) Groundwater extraction costs (B)

According to the *Report of Investigation to Groundwater Pollution in Key Sections of the Songhua River (Jilin City)* (2008), groundwater extraction mainly concentrates in the lower reaches of Mangniu River, Jiuzhan Industrial Zone and Hadawa Industrial Zone where there are 26, 19 and 12 pumping wells respectively. The calculation process of groundwater extraction costs and the results are as shown in Table 6.2. The extraction of 3.0057×10^8 m^3 of groundwater costs 6.0094×10^7 yuan.

(3) Compensation for resource development (Z_b)

Taxes are imposed on the development of non-renewable geological resources and used to support the development of alternative resources to ensure sustainable and balanced social and economic development. The income is referred to as compensation for mineral resources development and calculated using the following formula:

$$Z_b = F \times Q \quad (6.3)$$

Wherein Z_b represents compensation and Q output; F is tax rate.

Table 6.1 Quantity of groundwater pollution

Section	$S_{polluted\ aquifers}$ (km^2)	μ	$H_{aquifer}$ (m)	$Q_{pollution}$ (10^4 m^3)
1	15.114	0.14	5,041.804	15.114
2	16.035	0.16	6,738.052	16.035
3	11.123	0.17	2,418.987	11.123
4	16.863	0.18	2,916.116	16.863
5	25.380	0.2	12,701.17	25.380
Total				30,057.71

6.1 Aggregate Value of Groundwater Resources

Table 6.2 Groundwater extraction costs

Section	Number of wells	Height of aqueous layer (m)	Height of vadose zone (m)	Costs for drilling materials Drilling costs (yuan)	Costs for well materials Precipitation tubes (yuan)	Solid tubes (yuan)	Filter tubes (yuan)	Fillers (yuan)	Pumping costs Pumps	Pumping stations	Pumping operations	Labor
Mangniu River	26	15.00	1.90	183,953.3	1,981.2	754.6074	9,750	101,036	1,264,133	130,000	25,623,000	48,000
Hadawan Industrial Zone	12	10.00	3.50	71,721.97	914.4	640.8335	3,000	31,088	583,446	60,000	11,826,000	48,000
Jiuzhan Industrial Zone	19	23.00	2.71	188,456.2	1,447.8	784.1835	10,925	113,212.1	923,789.5	95,000	18,724,500	48,000
Note [a]		Data collected from 87 boreholes	Data collected from 87 boreholes	323 yuan/m, with a borehole diameter of 400 mm	15.24 yuan/m for precipitation tubes which are 5 m long each, made up of DN300 pre-stressed reinforced concrete tubes	15.24 yuan/m for DN300 pre-stressed reinforced concrete tubes whose length is the same as the height of vadose zone, using	25 yuan/m for Q2359 (A3) filter tubes whose length is same as height of aquifers	30 yuan/m³ for filter materials	48,620.5 yuan per 300QJ@230-82/4-45 submersible pump	About 5,000 yuan per simple pumping station	Motor power of 45 kWh, electricity price 0.5 yuan/kWh, 24 h a day in 5 years	Monthly wage of 800 yuan per person in 5 years
Total	57			444,131.5	4,343.4	2,179.624	23,675	245,336.1	2,771,369	285,000	56,173,500	144,000

Note [a]Shanghai Municipal Engineering Institute. Manual for Water Supply and Drainage Design (Volume 10) [M]. China Architecture and Building Press, April 2004, 2nd Edition, Plumbing Estimates, P255–283

As there are no standards concerning the tax rate for the development of groundwater resources, the tax rate for groundwater resource development (f) is set to 0.02 yuan/m^3 with reference to the *Administrative Measures of Henan on the Compensation for the Development of Mineral Resources*. Hence, the compensation for resource development (Z_b) amounts to 6.0114 × 10^6 yuan (3.0057 × 10^8 m^3 × 0.02 yuan/m^3)

(4) Compensation for geological prospecting and exploitation (K_b)

The compensation is determined by the costs of geological prospecting and exploitation or investment in geological prospecting and exploitation. Generally, an accumulative approach is used, which takes the sum of the actual investment in individual projects as the compensation. The Formula is written as follows:

$$K_b = W_f + J_f + Z_f + S_f + Q_f \tag{6.4}$$

Wherein K_b is the investment in geological prospecting and exploitation; and W_f represents physical exploration expenses, J_f well costs, Z_f equipment expenses, S_f test expenses, and Q_f other expenses.

As groundwater resource in Jilin City have been proven, the investment in geological prospecting and exploitation is barely needed. At the same time, hydrogeological exploration is relatively deepened and essentially supported by public welfare investment which can be ignored herein. Therefore, $K_b = 0$.

(5) Compensation for ecological damage (H_b)

The damage to the ecological environment brought by exploitation of groundwater resources includes land occupation, water pollution, land subsidence, earth sinking and earth fissures associated with falling groundwater tables, and weakened ecological functions of plants, microorganisms and other components due to increased depth to water tables. The compensation for ecological damage caused by groundwater extraction is the sum of compensation for damages to all these ecosystem functions.

In the city of Shijiazhuang, groundwater extraction (1.3 × 10^8 m^3) involves a total area of about 1.35 × 10^4 m^2 (Liu 2007). It is analogized that approximately 3.1213 × 10^4 m^2 of land is occupied to extract 3.0057 × 10^8 m^3 of groundwater in the study area. Given a price of 780 yuan/m^2 for level-1 industrial land in Jilin City, the costs associated with land occupation will be 2.4346 × 10^7 yuan.

The losses related to pollution caused by groundwater extraction can be avoided by taking appropriate measures.

According to existing literature and hydrogeological survey reports, the geological environment in the study area is free from such problems as cones of depression, land subsidence, earth sinking and earth fissures. The study area enjoys a large modulus of recoverable groundwater resources, among the most resource-rich areas of the province and has access to a wealth of recharge from the Songhua River for extraction. Under the current extraction conditions, the groundwater table does not show a downward trend. Given the current degree of exploitation, the study area has a considerable potential for groundwater extraction. Hence, groundwater exploitation in the study area has little impact on regional ecological functions associated with plants, microorganisms and so on.

The compensation for ecological damage (H_b) is 2.4346×10^7 yuan.

Given the equation $P_b = H_b + K_b + Z_b$, the compensated value (P_b) is 3.03574×10^7 yuan.

In 2005, the price of water for domestic use in Jilin City (L) was 1.0 yuan/m³. By inputting the obtained values of B, Q, H_b, K_b, and Z_b into Formula (6.1), the aggregate value of groundwater resources (k) can be calculated as follows:

$$k = (Q \times L - B) + (H_b + K_b + Z_b) = (3.0057 \times 10^8 \, m^3 \times 1.0 \, yuan/m^3 - 6.0094 \times 10^7 m^3)$$
$$+ (2.4346 \times 10^7 \, yuan + 0 \, yuan + 6.0114 \times 10^6 \, yuan) = 2.7083 \times 10^8 \, yuan.$$

6.2 Economic Losses of Groundwater Pollution

The contaminated groundwater may contain heavy metals, organic compounds and other toxic substances. According to the "concentration-value curve" proposed by L. D. James, the losses (S) caused by a specific contaminant in water bodies may be expressed as:

$$S = k \times R \qquad (6.5)$$

Wherein S represents the economic losses of specific water pollutant; k indicates the aggregate value of groundwater resources. R is the loss rate of the specific pollutant with C concentration, which refers to the ratio of economic losses to aggregate value of water bodies. It is calculated as follows:

$$R = \frac{S}{k} = \frac{1}{1 + a \times \exp(-b \times c)} \qquad (6.6)$$

Wherein a and b are parameters to be estimated, which are determined by contaminant characteristics; C_0 denotes the critical pollutant concentration to cause losses. Herein, a new variable is introduced $X = C/C_0$, then X is a dimensionless number. The above formula is rewritten as:

$$R = \frac{1}{1 + a \times \exp(-b \times C_0 \times X)} = \frac{1}{1 + A \times \exp(-B \times X)} \qquad (6.7)$$

Parameters A_{ij} and B_{ij} are estimated based on the measured data, using the following formula:

$$\begin{cases} A_{ij} = 99^{(x_{ij}+1)/(x_{ij}-1)} \\ B_{ij} = 2\log 99/(x_{ij}-1) \end{cases} \tag{6.8}$$

The comprehensive loss rate of multiple contaminants is not equal to the algebraic sum of the loss rate of specific pollutants, but rather a set. In the presence of n independent contaminants, the comprehensive loss rate $R_i^{(n)}$ is calculated as follows:

$$R_i^{(n)} = R_i^{(n-1)} + (1 - R_i^{(n-1)})R_{in} \tag{6.9}$$

The critical concentrations of contaminants are referred to as the Class III levels of the Drinking Water Quality Standards (2006). The A, B and R values of typical pollutants are calculated according to Formula 6.7 and Formulated 6.8, with results summarized in Table 6.3. Then, the comprehensive loss rate is calculated $R_i^{(10)} = 77.29\%$.

Given an aggregate value of groundwater resources obtained in Sect. 6.1, the direct economic losses caused by groundwater pollution can be calculated as follows:

$$S = k \times R = 2.7083 \times 10^8 \text{ yuan} \times 77.29\% = 2.0932 \times 10^8 \text{ yuan}$$

The total economic losses (S_{total}) caused by groundwater pollution encompass direct economic losses identified through concentration-value calculations and indirect economic losses associated with the irreparable damage to human health and groundwater aquifer. It is generally believed at home and abroad that the indirect economic losses are four to five times of the direct economic losses (Tan and Guan, 2004). Herein, a coefficient of 4.5 is taken, then the total economic losses are:

$$\begin{aligned} S_{total} &= S + S_{indirect} = 2.0932 \times 10^8 \text{ yuan} + 4.5 \times 2.0932 \times 10^8 \text{ yuan} \\ &= 11.5126 \times 10^8 \text{ yuan} \end{aligned}$$

In 2004, the environmental cost of pollution was equivalent to 3% of China's gross domestic product (GDP), and the proportion stayed around 3.8% in 2008 and 2009 and approximately doubled to 5–6% in 2011 (Wang 2012). According to the *Environmental Cost of Pollution in China* of the World Bank released in 2007, China's annual economic losses caused by pollution amount to 100 billion US dollars, accounting for 5.8% of the GDP, and the proportion will rise to 9% in 2009. As far as Jinlin City is concerned, the cost of groundwater pollution in the study area represented 1.8% of the GDP which was 629.7 billion yuan in 2005. Groundwater pollution took up a relatively large proportion in the environmental cost, compared with reported. This implies that the groundwater crisis in the study area increasingly dampens economic development, so it is urgent and necessary to control groundwater pollution risk and prevent groundwater pollution.

6.2 Economic Losses of Groundwater Pollution

Table 6.3 A, B and R values of typical pollutants

Indicator	COD	TDS	NH_4-N	NO_3-N	SO_4^{2-}	TH	Hg	Benzene	BaP	Carbon tetrachloride
C (mg/L)	4.26	1,470.01	31.112	112.9	454.3	657.54	0.0068	0.0497	0.000314	0.00551
C_0 (mg/L)	3	1,000	0.5	20	250	450	0.001	0.01	0.00001	0.002
X	1.42	1.47001	62.224	5.645	1.8172	1.4612	6.8	4.97	31.4	2.755
A	3.15E + 11	3.07E + 10	115.03	715.973	7.58 E + 06	4.46E + 10	482.82	1,002.29	133.95	18,615.01
B	9.5030	8.4919	0.0652	0.8593	4.8841	8.6541	0.6882	1.0054	0.1313	2.2742
R	2.30E-06	8.59E-06	0.3343	0.1515	9.43E-04	6.95E-06	0.1824	0.1286	0.3154	0.0275

References

Jilin Hydrogeological Investigation Report. Environmental Monitoring Center of Jilin Province; 2008.

Liu CL. Economic evaluation of geological environmental risk of cities. Shijiazhuang: Chinese Academy of Geological Sciences; 2007.

Report of investigation to groundwater pollution in key sections of the Songhua River (Jilin City). Environmental Monitoring Center of Jilin Province; 2008.

Tan CL, Guan H. Environmental economics. Beijing: Science Press; 2004.

Wang YQ. "Twelfth Five-Year" pollution reduction and green transformation proposals. Environ Prot 2012;(19):16.17.

Chapter 7
Groundwater Pollution Control Risk from the Perspective of Industrial Economics

Abstract When the wastewater discharge in the process of economic growth exceeds the groundwater capacity, the groundwater pollution may occur. To avoid the whole social-economic environment system collapsing, it's fundamentally necessary to adjust the industrial structure. This chapter takes Jilin City for example and proposes the specific measures of groundwater pollution control from the perspective of adjusting the industrial structure according to the local groundwater pollution condition, features of industrial economy and pollution features of different industries. The results show that industrial pollution sources were found to exert the most significant impact on the groundwater pollution. The features of Jilin's industrial development included obvious resource-dependency, distinct leading characteristic of heavy chemical industry and relatively low technical level of the industry. The three characteristics commonly gave rise to the local groundwater pollution condition. Hence, it's necessary to take steps to control the groundwater pollution, such as strengthening the structural adjustment of different industries, improving the efficiency of different industries and enhancing to build the public platform and infrastructure. The steps have been proved to be effective in Jilin city to some extent, and need to be further promoted for solving the groundwater pollution problem.

Keywords Industrial structure adjustment · Groundwater pollution
Pollutant characteristics

Agriculture, industry, and urban life cause groundwater pollution in the process of economic and social development. According to the results of groundwater pollution risk assessment in Sect. 5.4, the high and relatively high risk of groundwater pollution can be mainly attributed to industrial point sources (sewage outfalls), linear sources (Jiuzhan open channel) and non-point sources (industrial sites), particularly chemical plant, chemical fiber plant, pesticide plant, thermal power plant, calcium carbide plant, and cement plant. Given this, this chapter focuses the analysis on industry.

7.1 Thinking on Industrial Restructuring and Its Priorities

In view of the problems existing in the industrial development of Jilin City, the industrial structure can be optimized in three aspects for the purpose of groundwater pollution risk control: (a) transition from high-pollution heavy chemical industry to other low-pollution industries, which considers both the incremental capacity of emerging industries and the existing capacity of traditional industries; (b) technological progress and efficiency improvement of industrial sectors, which aims to reduce the emission intensity by improving production technologies, equipment and processes; (3) construction of infrastructure and public platforms, covering sewage treatment facilities, technologies, equipment and platforms that facilitate pollution control. Among them, the first and second aspects address pollution sources to cut emissions, while the third aspect controls pollution by digesting pollutants. As the high-pollution heavy chemical industry takes up a large proportion and many industries have lower technical efficiency than the national average, there is a large potential for structural adjustment and efficiency improvement in Jilin City, which is therefore considered as the priorities of industrial restructuring. In comparison, the time-taking structural adjustment is a medium- and long-term strategy oriented to future development. Efficiency improvement is a medium- and short-term development strategy that requires additional equipment and technology investment. Infrastructure and public platform construction serves as a major measure for near-term pollution control, whose technical and capital conditions can be easily met.

In accordance with the thinking on industrial restructuring and its priorities, Jilin City has achieved some success in pollution control through a series of policy measures in recent years. It was identified as one of the first eight model cities for fiscal policy on energy conservation and emissions reduction by the Ministry of Finance (MOF) and the National Development and Reform Commission (NDRC) in June 2011. With focus on four major projects including circular transformation of chemical industrial parks, Jilin City fosters the systems of low carbon, green building, pollution control, clean transportation, renewable energy and intensive service, aimed at breakthrough progress in industrial restructuring, significant improvement of energy efficiency in key fields, drastic reduction of pollutant emissions, and more complete urban infrastructure. In 2011, a total of 37 key abatement projects were launched, covering COD (13), ammonia nitrogen (4), sulfur dioxide (16) and nitrogen oxide (4). Their emission reductions amounted to 5,490 tons, representing 132.3% of the 4,142 tons tasked by Jilin Province (Jilin Social and Economic Statistical Yearbook, 2012). By taking a variety of measures, the total discharge of industrial wastewater continued to decline from 2007 onwards and fell to 112.91 million tons in 2011, the lowest in recent years. The COD emissions also reached a record low of 15,596 tons in 2011, but the ammonia nitrogen emissions showed a rising trend in these years, indicating grim emissions reduction situation (Table 7.1).

7.2 Path for Industrial Restructuring 111

Table 7.1 Industrial wastewater discharge in Jilin City (2005–2011)

Year	Total (10,000 tons)	COD (tons)	Ammonia nitrogen (tons)
2005	18,693	28,625	2,426
2006	16,685	27,199	2,304
2007	18,293	27,952	342
2008	16,168	25,333	467
2010	15,037	31,192	529
2011	11,291	15,596	866

Source *Social and Economic Statistical Yearbooks of Jilin City* (2005–2011)

7.2 Path for Industrial Restructuring

7.2.1 Intensify Structural Adjustment of Industries of Different Types

Figure 7.1 shows the change in the output value structure of heavy and light industries in Jilin City during 2005–2011 and Table 7.2 provides statistics on industrial structure of Jilin City in 2005 and 2011. Through the adjustment of industrial and product structure, the obvious tilt to polluting heavy industries has been alleviated. From 2005 to 2011, the proportion of heavy industry in industrial output value decreased from 88.09 to 75.23%, a drop of 12.86% points, while that of light industry increased from 11.91 to 24.77%, nearly 1/4 of the total. Particularly, the percentage of industries with high pollution intensity fell markedly by 19.71% points from 66.48 to 46.77%. In specific, the decline reached 24.44% points from 57.27 to 32.83% in chemical and chemical product manufacturing industry and 2.20% points from 4.81 to 2.62% in chemical fiber manufacturing industry. Although there was an increase of 12.72% points from 4.17 to 16.88% in industries with relatively high pollution intensity (noticeably 8.84% points in

Fig. 7.1 Change in the percentage of heavy and light industries in Jilin City (%). *Source* Calculations based on the *Social and Economic Statistical Yearbook of Jilin City (2012)*. The figure is reprinted from (Huan et al. 2016), with permission from Research on development

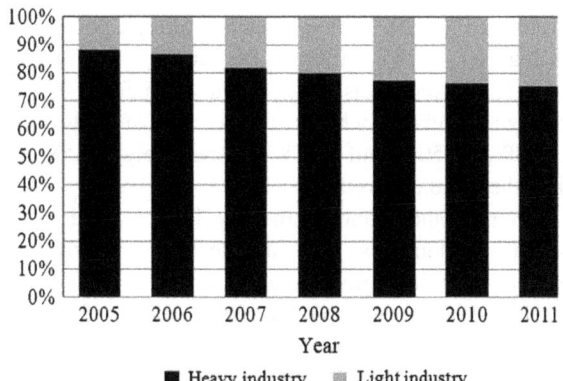

Table 7.2 Industrial structure comparison of Jilin City in 2005 and 2011

Industry	2005		2011	
	Total output value (10,000 yuan)	Percentage (%)	Total output value (10,000 yuan)	Percentage (%)
High pollution intensity	5,017,209	66.48	12,934,869	46.77
Relatively high pollution intensity	314,414	4.17	4,669,935	16.88
Relatively low pollution intensity	1,034,083	13.70	4,562,615	16.50
Low pollution intensity	1,163,252	15.41	5,306,177	19.18

Source Calculations based on the *Social and Economic Statistical Yearbook of Jilin City (2012)* and *China Statistical Yearbook 2012*. The table is reprinted from (Huan et al. 2016), with permission from research on development

agro-food processing industry), the percentage of industries high and relatively high pollution intensity came down by 6.99% points.

On the one hand, resource-consuming and high-pollution enterprises were required to rectify within the prescribed time, or shut down, relocated and converted, in order to vigorously cut backward production capacity. For example, JCGC has phased out 69 sets of devices since the 10th FYP period and another 13 sets including silicone device in 2006. Jilin City removed 680,000 tons of backward cement production capacity and 15,000 tons of viscose staple fiber in 2011, and 280,000 tons of backward production capacity of Jilin Chenming Paper Co., Ltd in 2012 amid further effort in technological transformation and backward production capacity elimination. In particular, Jilin Chenming Paper Co., Ltd was included in the 2012 special work under the national program for backward production capacity elimination. It took off out the 5,560 newsprint production line purchased from Finland in 1968 and the 200-ton and 300-ton chemical pulp production lines put into operation in 1980. With the exit of energy-consuming, poor-equipped unstable production lines, the COD emissions were curtailed by 89% year on year (website of the People's Government of Jilin City 2013). From the national point of view (Table 7.3), in 2011, the percentage of high-pollution-intensity industries remained as high as 46.77% in Jilin City, 26.44% points more than the national average (20.33%), which was a noticeable drop from 45.40% points in 2005. The percentages of relatively high, relatively low and low pollution intensity industries read 16.88%, 16.50% and 19.18% respectively, 13.55, 5.26 and 6.45% points below than the national average.

On the other hand, Jilin City implemented a series of policies in favor of high-tech industries such as electronic information and medicine. It placed emphasis on strategic emerging industries with high knowledge and technology content, low resource consumption and great growth potential, and effectively promoted the development of modern high-tech industries by integrating science and technology. In 2011, the Work Group for the Promotion of Strategic Emerging

7.2 Path for Industrial Restructuring

Table 7.3 Comparison of Jilin City and China's industrial structure in 2011

Industry	Jilin City		National	
	Total output value (10,000 yuan)	Percentage (%)	Total output value (10,000 yuan)	Percentage (%)
High pollution intensity	12,934,869	46.77	171,645	20.33
Relatively high pollution intensity	4,669,935	16.88	256,957	30.44
Relatively low pollution intensity	4,562,615	16.50	183,676	21.76
Low pollution intensity	5,306,177	19.18	216,412	25.63

Source Calculations based on the *Social and Economic Statistical Yearbook of Jilin City (2012)* and *China Statistical Yearbook 2012*. The table is reprinted from (Huan et al. 2016), with permission from research on development

Industries was set up, covering new materials, advanced equipment manufacturing, biomedicine, electronic information, bio-chemicals, energy conservation and environmental protection, new energy, new energy vehicles, culture and tourism. Driven by a variety of policy measures, these strategic emerging industries gained a strong momentum of development. As of June 2011, there were 585 enterprises of strategic emerging industries (apart from cultural and tourism industries) in the city, of which 211 were above the designated scale (Yang 2011). In 2011, the strategic and emerging industries yielded an output value of 56.9 billion yuan, of which the added output value increased by 42% year on year to 16.8 billion yuan, forming an important part of industrial economic growth (Social and Economic Statistical Yearbook of Jilin City 2012). At present, the industries of new materials, bio-medicine, bio-chemicals, electronic information, advanced equipment manufacturing, energy conservation and environmental protection have reached a certain scale with appropriate foundation. As a highlight of strategic emerging industries, the carbon fiber industry has leaped forward with the commercialization of key projects of Jilin Chemical Fiber Co., Ltd and Jiyan High-Tech Fibers Co., Ltd. In 2011, the production capacity of carbon fiber precursor, carbon fiber and carbon fiber attained 5,400 tons, 818 tons and 150 tons respectively, making the city a domestic leader of the industry. There was also a rise in the output value percentage of the industries of manufacturing transportation equipment, special equipment, and electrical machinery and equipment, compared with 2005. The percentage increased by 2.33% points for electrical machinery and equipment manufacturing industry and to 4.59% for transportation equipment manufacturing industry.

7.2.2 Push for Efficiency Improvements in Various Industrial Sectors

Despite a solid foundation of traditional industries in Jilin City, technological innovation to drive industrial upgrading is urgent in view of the prevalent equipment aging, backward technologies and uncompetitive products. To this end, the city has made vigorous efforts targeted at technological progress of enterprises, including the implementation of science and technology projects, establishment of university science and technology parks, and construction of public service platforms for technology transfer and property rights trading. Meanwhile, it continued to increase investment in technology research and development, and intensify technological innovation and transformation, so that technological progress can help to upgrade product quality and stimulate new product development. In 2011, totally 1,800 the science and technology projects at the municipal level and above were carried out, involving an investment of 91.5 million yuan, which cultivated 22 enterprise-based technology centers and 45 scientific and technological innovation teams.

As shown in Table 7.4, in 2011, there was an obvious improvement of productivity in industries with low and relatively low pollution intensity compared with 2005. Among the 18 fastest productivity advancers, 6 industrial sectors have low pollution intensity, 5 relatively low pollution intensity, and 3 relatively high pollution intensity. The improvement was limited in the industries of chemical and chemical product manufacturing, chemical fiber manufacturing, and pharmaceutical manufacturing. These three industrial sectors which occupy an important position in Jilin City ranked 20, 27 and 21 respectively.

Considering groundwater pollution control, industrial efficiency improvements should be focused on two points. (1) In view of technological backwardness, equipment aging, and high pollution of enterprises, production processes and equipment with low efficiency of resources and energy and large emissions of "three wastes" should be removed, while new technologies, processes, and facilities applied to improve productivity and develop products. Especially, technological innovation will be carried out in a number of well-founded, large-scale, prospective backbone enterprises. This will contribute to energy conservation and emissions reduction and efficient waste reuse and comprehensive utilization while improving product quality and market competitiveness. For example, PetroChina Jilin cuts emissions year by year and controls pollutant concentrations far below the national standards through green and eco-friendly equipment and refined production process and technology. Among them, the desulfurization and denitrification of No. 4 boiler of Second Power Plant reduces 1,200 tons of nitrogen oxides per year. (2) Industrial chain collaboration should be optimized by linking technologies, products and processes of different enterprises, so as to improve the overall efficiency of the industrial chain. The construction of industrial agglomerations can be an effective way. For example, PetroChina Jilin Low-Carbon Economy Demonstration Park follows the direction of raw material diversification, industrial development

7.2 Path for Industrial Restructuring

Table 7.4 TFP comparison of Jilin City in 2005 and 2011 (by added value)

Industrial sectors	2005	2011	2011/2005
Printing and recording media	−0.27	2.18	–
Gas production and supply	−0.14	4.43	–
Oil and gas extraction	1.60	17.86	11.15
Wood processing and wood, bamboo, rattan, palm and grass products	0.94	5.41	5.77
Non-metallic mining	0.94	5.00	5.33
Instrumentation and cultural and office machinery manufacturing	0.96	4.91	5.12
Furniture manufacturing	1.07	4.74	4.44
Electrical machinery and equipment manufacturing	0.82	3.49	4.26
Coal mining and washing industry	0.54	2.17	4.01
Non-metallic mineral products	0.85	3.27	3.86
Beverage manufacturing	1.76	6.68	3.79
Textiles	0.93	3.14	3.36
Paper and paper products	0.70	1.99	2.86
Transportation equipment manufacturing	1.12	2.83	2.53
Special equipment manufacturing	0.98	2.30	2.36
Textile apparel, footwear and hat manufacturing	0.53	1.21	2.30
General equipment manufacturing	1.60	3.46	2.17
Agricultural and sideline food processing	2.16	4.63	2.14
Handicrafts and other manufacturing	1.60	3.18	1.99
Chemical and chemical product manufacturing	1.57	3.07	1.95
Pharmaceutical manufacturing	1.12	2.15	1.91
Ferrous metal smelting and calendering	2.35	3.98	1.69
Food manufacturing	1.78	2.95	1.66
Electricity and heat production and supply	1.79	2.82	1.58
Rubber products	2.18	2.73	1.25
Ferrous metal mining	3.77	4.60	1.22
Chemical fiber manufacturing	1.64	1.97	1.20
Non-ferrous metal smelting and calendering	2.60	3.10	1.19
Non-ferrous metal mining industry	2.25	2.41	1.07
Water production and supply	0.74	0.74	1.01
Communications equipment, computers and other electronic equipment	1.55	1.47	0.95
Leather, fur, feathers (down) and their products	1.18	0.93	0.78
Metal products	3.58	2.60	0.73

Source Calculations based on the *Social and Economic Statistical Yearbook of Jilin City (2012)* and *Social and Economic Statistical Yearbook of Jilin City (2016)*

extension, and product processing refinement, and creates mutually related and supportive industrial systems of fine chemicals, chemical new materials, and renewable energy to advance the transition from traditional chemical industry to modern ecological chemical industry. In terms of wastewater reuse, the demonstration park establishes a three-level system that effectively controls the discharge of a variety of wastewater. In addition, water recycling becomes better by substantially increasing the concentration ratio of circulating water and strengthening the management of water-cooling equipment.

As mentioned above, Jilin City has achieved remarkable results in technological transformation and innovation in the industrial fields. The TFP comparison, as shown in Table 7.5, indicates that in 2011, more than half of Jilin City's industrial sectors had productivity higher the national average, noticeably chemical and chemical product manufacturing, beverage manufacturing, food manufacturing, and metal products. In other industrial sectors, however, the productivity was below the national average, such as non-ferrous metal mining, chemical fiber manufacturing, pharmaceutical manufacturing, textile apparel, footwear and hat manufacturing, communications equipment, computers and other equipment manufacturing, and leather, fur, feather (down) and their products.

7.2.3 Advance the Construction of Public Platforms and Infrastructure

The construction and capacity of wastewater treatment facilities in Jilin City in different years is compared (Table 7.6). The wastewater treatment capacity gradually expanded with the investment in infrastructure construction, which plays a huge role in improving the local ecological environment and promoting economic and social sustainable development. In 2011, there were 126 sets of wastewater treatment facilities in the city, with a total daily capacity of 147 tons and an annual capacity of 46 million tons. Among them, as an important component of Songhua River comprehensive management project, the municipal wastewater treatment plant was put into operation in 2007. The plant involving an investment of 620 million yuan embraces domestically leading technical equipment, process, and technology. With a daily processing capacity of 300,000 tons, the plant can effective treat more than 90% of domestic and industrial wastewater in the city. However, with urban construction and economic development, wastewater discharge continues to increase beyond 600,000 m^3/d, the capacity of Jilin sewage treatment pipeline project completed during the 11th FYP period. To this end, Jilin City kicked off the Phase II project of municipal wastewater treatment plant whose ultimate designed capacity is 300,000 m^3/d and near-term capacity is 150,000 m^3/d (2015). In 2011, combined with the water pollution control plan for the Songhua River, the city proposed to invest 1.45 billion yuan in 12 projects of urban sewage treatment and facilities and 40 million yuan in the comprehensive improvement

7.2 Path for Industrial Restructuring

Table 7.5 TFP comparison of Jilin City and China in 2011 (by output value)

Industrial sectors	China	Jilin City	Jilin City/ China
Oil and gas extraction	9.57	22.92	2.40
Non-metallic mining	13.10	21.92	1.67
Beverage manufacturing	11.43	18.96	1.66
Furniture manufacturing	9.88	15.30	1.55
Chemical and chemical product manufacturing	14.78	20.27	1.37
Handicrafts and other manufacturing	10.80	14.79	1.37
Food manufacturing	12.49	15.33	1.23
Instrumentation and cultural and office machinery manufacturing	9.43	11.01	1.17
Metal products	11.55	13.44	1.16
Electrical machinery and equipment manufacturing	11.67	12.56	1.08
Gas production and supply	13.81	14.84	1.07
Textiles	10.27	11.02	1.07
Agricultural and sideline food processing	17.84	18.47	1.04
Non-metallic mineral products	11.11	11.48	1.03
General equipment manufacturing	11.50	11.67	1.01
Wood processing and wood, bamboo, rattan, palm and grass products	13.96	14.02	1.00
Paper and paper products	10.50	10.51	1.00
Coal mining and washing	7.65	7.15	0.93
Ferrous metal smelting and calendering	17.02	15.41	0.91
Rubber products	11.69	10.54	0.90
Non-ferrous metal mining	13.48	11.34	0.84
Ferrous metal mining	14.13	11.82	0.84
Printing and recording media	8.83	7.04	0.80
Textile apparel, footwear and hat manufacturing	8.59	6.58	0.77
Special equipment manufacturing	10.43	7.88	0.76
Water production and supply	3.04	2.25	0.74
Leather, fur, feather (down) and their products	9.09	6.55	0.72
Transportation equipment manufacturing	12.42	8.90	0.72
Pharmaceutical manufacturing	10.86	7.12	0.66
Chemical fiber manufacturing	14.80	8.67	0.59
Electricity and heat production and supply	11.79	5.18	0.44
Communications equipment, computers and others	11.60	4.42	0.38
Non-ferrous metal smelting and calendering	18.38	5.03	0.27

Source Calculations based on the *Social and Economic Statistical Yearbook of Jilin City (2012)* and *China Statistical Yearbook 2012*. The table is reprinted from (Huan et al. 2016), with permission from research on development

Table 7.6 Construction and capacity of wastewater treatment facilities in Jilin City

Year	Number of wastewater treatment facilities	Wastewater treatment capacity (10,000 tons/day)	Wastewater inflow of municipal wastewater treatment plant (10,000 tons)
2005	142	80	3,542
2006	147	81	4,246
2007	145	88	2,619
2008	113	87	2,636
2010	112	83	2,509
2011	126	147	4,600

Source Social and Economic Statistical Yearbooks of Jilin City (2005–2011)

project for regional water environment. In addition, the project of provincial hazardous waste treatment center, launched in 2008, was completed and ready for trial operation. Involving an investment of about 170 million yuan, the provincial hazardous waste treatment center is capable of landfilling 35,110 tons of hazardous waste annually.

In the meantime, Jilin City strictly required enterprises to introduce environmentally friendly equipment. For example, PetroChina Jilin wastewater treatment plant, one of the first comprehensive sewage treatment plants in the country, has treated 1.5 billion m^3 of sewage and removed more than 800,000 tons of pollutants since 2010 (China Chemical Industry News 2010), playing an important role in mitigating the pollution of the Songhua River Basin. In recent years, the plant conducted technical transformation of the 70,000 m^3 buffer pool to serve for acid hydrolysis in the production state. As a result, sewage can be treated more deeply and precisely, which substantially increases the COD removal rate and further curtail the COD discharge. The 300,000-tons aniline plant in Jilin Economic and Technological Development Zone completely meets the emissions standards by installing the column washer worth 200 million yuan that removes organic matter and nitride generated in the production process. The wastewater treatment works launched in recent years cover Shulan Synthetic Pharmaceutical Co., Ltd, Jilin Chenming Paper Co., Ltd. FAW Jilin Automobile Co., Ltd., Jilin Chemical Fiber Co., Ltd. (technological transformation for production wastewater reuse and comprehensive utilization), and Jilin Jien Nickel Industry Co., Ltd. (heavy metal treatment). In the process of industrial restructuring, infrastructure construction has been combined with industrial layout adjustment. Planning and guidance has been provided to relocate high-tech, light-pollution enterprises in high-tech industrial zones and heavy-pollution enterprises in the economic and technological development zones, so that pollutants can be treated in a centralized manner. In Jilin Economic and Technological Development Zone, for example, the wastewater treatment plant offers services to such enterprises as GE Polycarbonate Company, Cargill China, Jilin Chemical Fiber Co., Ltd, and Jilin Fuel Ethanol Co., Ltd.

7.3 Summary and Implication

(1) Groundwater pollution risk is mainly derived from the characteristics of industrial structure.

In the economic and environmental systems, economic growth and pollution emissions are inseparable and mutually constrained in the economic and social development. Groundwater pollution in a region depends to a large extent on its industrial economic characteristics. This has been clearly demonstrated by the environmental Kuznets curve proven by the experience of the developed countries. More specifically, the environment first deteriorated due to a sharp increase in resource consumption and pollutant emissions with rapid development of heavy chemical industry in the process of industrialization. Then, the environmental situation turned better in the stage of low consumption, emissions and pollution owning to the rise of technology-intensive industries and services. In this sense, pollution characteristics vary in stages or regions because of significant differences in industrial structures. In addition, the dominant industry in the region corresponds, to a large extent, to the level of regional economic development. Therefore, groundwater pollution control is interrelated and integrated with industrial restructuring. Only by clarifying the mechanism of interaction among groundwater pollution, industrial structure characteristics and industrial pollution characteristics, can it possible to accurately grasp the groundwater pollution and develop targeted control measures. The blind control on the quantity or concentration of pollutants decoupled from industrial structure would be hardly effective for groundwater pollution control.

(2) The key to groundwater pollution control lies in the implementation of differentiated policies on industrial structure.

Industrial restructuring is crucial to groundwater pollution control. The focal point is to adapt industrial structural policies to the local conditions of different regions, so as to address the pollution problems specific to industries. While drawing on the experience of advanced countries and regions, the simply replication or imitation of practice is difficult to take effect because industrial structure policies should be different considering different environmental elements, development stages, and functional positions.

First, industrial structure has certain path dependence, so industrial restructuring policies must reflect regional realities and respect objective facts. Jilin City, for example, as a traditional heavy chemical city, has a solid historical foundation in chemical, metallurgy, and mining industries. Even if faced with high pollutant emissions, the city should take appropriate measures under the premise of maintaining economic stability and orderly industrial convergence, and must properly deal with the relationship between efficiency improvement of traditional industries and cultivation of emerging industries.

Second, the industries that lead to severe groundwater pollution may also be different. Where large wastewater discharge pollutes the groundwater, such as beverage manufacturing, food manufacturing, and coal mining, the focus of structural adjustment is to increase water reuse and implement water-saving projects. Where pollutants are severe and difficult to control, such as pharmaceutical manufacturing, metal products, and waste recycling, the focus should be put on technological progress and enhanced regulation for pollution prevention and control.

References

China Chemical Industry News. Three-decade operation of Jilin petrochemical sewage treatment plant. 18 Oct 2010. http://www.jlsina.com/news/jilinshi/2011-1-28/10038.shtml.

Editorial Board of Social and Economic Statistical Yearbook of Jilin City. Social and economic statistical yearbook of Jilin City. Beijing: China Statistics Press; 2012.

Huan H, Li J, Li MX. Research on industrial structure adjustment of resource-based City from the perspective of groundwater pollution control: a case study in Jilin City. Res Dev. 2016;6:106–12.

Website of the People's Government of Jilin City. Chenming Paper Co., Ltd passes examination of backward production capacity elimination. 17 May 2013. http://www.chinairn.com/news/20130520/115904789.html.

Yang BQ. Analysis of current strategic emerging industries of Jilin City. 2011; (12). http://www.jlst.gov.cn/JLKJJ/DZKWDetail.jsp?id=78.

Chapter 8
Conclusions

Abstract With a case from Jilin section of the Songhua River, this study applies the overlay index method to assess the shallow groundwater pollution risk and the "concentration-loss" model to calculate economic losses of groundwater pollution, covering a variety of pollutants from different sources, and on this basis, put forward options to prevent and control groundwater pollution from the perspective of industrial economics. The chapter summarizes the conclusions obtained by the study.

Keywords Groundwater pollution characteristics
Groundwater pollution risk mapping · Risk control · Industrial economics

8.1 Groundwater Pollution Characteristics

Groundwater pollution in the study area can be categorized into industrial source pollution, agricultural source pollution and domestic source pollution. It can be caused by point sources (industrial and domestic sewage outlets, industrial solid waste, domestic garbage and manure), linear sources (Jiuzhan open channel), and non-point sources (fertilizers and pesticides in paddy fields, dry fields and vegetable fields). The chemical pollutants include COD, NO_3-N, NH_4-N, TFe, Mn, fluoride, TDS, SO_4^{2-}, Al, heavy metals (Hg, Zn and Cr), organic matters (volatile phenols, BaP, carbon tetrachloride, benzene and chlorobenzene), and the like.

8.2 Groundwater Pollution Risk Assessment Results

There is a low possibility of groundwater pollution overall as 67.61% of the study area has medium or lower risk of groundwater pollution. In contrast, relatively high and high risk of groundwater pollution is found in 32.39% of the study area, mainly due to pollution loads of industrial point sources (sewage outlets), linear sources

(Jiuzhan open channel) and non-point sources (industrial areas). The dominant factors of influence on groundwater pollution vary in these high-risk areas, such as groundwater function in the first-order terrace of Jiuzhan, south bank of Mangniu River, vicinity of Gujiazi in (western) Jiangbei District, and first-order terrace of Jinzhu District and intrinsic groundwater vulnerability in the south of Mangniu River, Gujiazi and Bajiazi in (western) Jiangbei District, and second-order terrace of Jinzhu District.

The reliability of groundwater pollution risk assessment results is verified using the level difference method. The reasonable area accounted for 64.45% of the study area, and 70.8% of sampling sites are located in the reasonable area. Overall, the risk assessment methodology is reasonable and applicable to the study area.

8.3 Aggregate Groundwater Value and Economic Losses Caused by Groundwater Pollution

The aggregate value of groundwater totals 2.7083×10^8 yuan, composed of resource base value (2.4048×10^8 yuan) and compensated value (0.3035×10^8 yuan). According to the concentration—value curve, the rate of loss caused by pollutants is calculated to reach 77.29%. Therefore, the direct economic losses caused by groundwater pollution are up to 2.0932×10^8 yuan and the total economic losses amount to 11.5126×10^8 yuan, accounting for 1.8% of the gross domestic product. As the groundwater crisis in the study area increasingly restricts economic development, it is urgent and necessary to control groundwater pollution risk and prevent groundwater pollution.

8.4 Groundwater Pollution Risk Control from the Perspective of Industrial Economics

Economic growth and pollution emissions are inseparable and mutually constrained in the economic and social development. The economic and environmental systems will be in a state of high risk when wastewater discharge in the process of economic growth exceeds groundwater capacity. Lowering the risk can be achieved by adjusting the industrial structure and improving the efficiency. In light of different pollution characteristics, industrial sectors can be divided into four categories according to pollution intensity of wastewater (yuan/10,000 yuan). High pollution intensity is mainly seen in chemical industry with huge water consumption. Industries with relatively high intensity are dispersed, including chemical industry, water-consuming light industry and equipment manufacturing industry characterized by difficult pollution treatment. The pollution intensity is medium and low in industries with a short industrial chain and independent of other raw materials. The characteristics of the industrial economy of Jilin city include prominent reliance on sources, clear dominance of heavy chemical industry, relatively low technical level and wide spatial variance.

According to the above analysis, the groundwater pollution risk is mainly derived from the characteristics of industrial structure. The major measures for industrial restructuring should take into account groundwater pollution risk control, such as restructuring industries of different types, improving the efficiency of industrial sectors, and advancing public platform and infrastructure construction. The key to the prevention and control of groundwater pollution lies in the implementation of differentiated structural policies which address the problems specific to industries according to the local conditions of different regions.

Lightning Source UK Ltd.
Milton Keynes UK
UKHW02n1143040518
322118UK00002B/60/P